Packet Radio Primer

Dave Coomber, G8UYZ
and
Martyn Croft, G8NZU

Radio Society of Great Britain

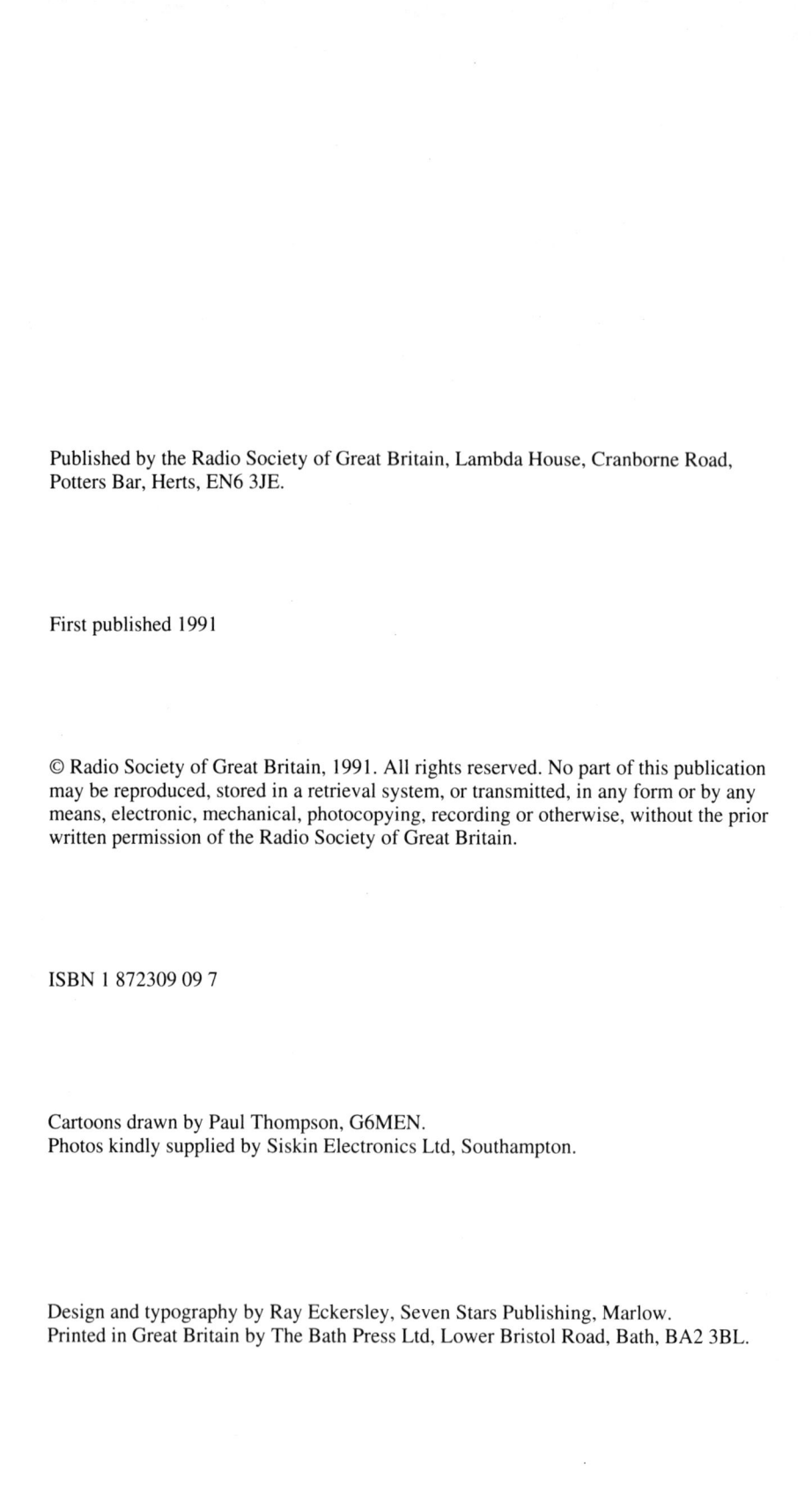

Published by the Radio Society of Great Britain, Lambda House, Cranborne Road, Potters Bar, Herts, EN6 3JE.

First published 1991

© Radio Society of Great Britain, 1991. All rights reserved. No part of this publication may be reproduced, stored in a retrieval system, or transmitted, in any form or by any means, electronic, mechanical, photocopying, recording or otherwise, without the prior written permission of the Radio Society of Great Britain.

ISBN 1 872309 09 7

Cartoons drawn by Paul Thompson, G6MEN.
Photos kindly supplied by Siskin Electronics Ltd, Southampton.

Design and typography by Ray Eckersley, Seven Stars Publishing, Marlow.
Printed in Great Britain by The Bath Press Ltd, Lower Bristol Road, Bath, BA2 3BL.

Contents

Preface	v
1. Introduction	1
2. OK, what is packet radio?	5
3. Getting started	13
The radio • The TNC • The terminal • Computer requirements • Software requirements • Data links	
4. Setting up the station	23
The terminal to the TNC • An example connection • The TNC to the radio • Interference • Setting up the TNC • The switch-on	
5. Getting on the air	35
Essential commands (to get you up and running) • Definitely useful commands • A few tweaks (those minor adjustments) • Some basic do's and don'ts • A first contact • Hopping around	
6. The packet postman	49
The mailbox • Example session on a mailbox • Downloading files • A personal mailbox • Accessing another PMS • TNC ROM version 1.1.6 • Using the PMS as the sysop • A few other useful commands • A sample of using my own PMS	
7. Packet protocols	65
What's in a packet? • A note on nodes • DX PacketClusters	
Epilogue	79
Appendix 1. TNC-2 and clone command list	81
Appendix 2. Packet mailbox commands	89
The WA7MBL BBS • The W0RLI mailbox • The AA4RE mailbox • The G4YFB BBS mailbox software • The G1NNA mailbox software • The FBB BBS software	
Appendix 3. A short guide to communication cables	115
Appendix 4. Some PC software	119
YAPP, v2.0 • ProComm, v2.41 • A brief look at Paket, v4.0	
Appendix 5. A short glossary	129
Appendix 6. Guidelines for the use of the packet radio network	131
Index	133

Preface

Maybe one of the reasons why amateur radio has endured as a hobby is that radio and the peripheral aspects are in constant change; there is always something new around the corner. A first introduction to amateur radio might well stress that upon gaining the licence, you are relatively free to pick and choose those aspects of the hobby which interest you most. And to ensure that you have plenty of choice, new developments in every facet of radio are frequently reported.

For instance, who would ever have dreamt, even 10 years ago, of using a portable 2-metre transceiver to link a home computer to an international information network? Yet electronic mail linking like-minded amateur radio enthusiasts is now a reality, with more to come. Written for the beginner in packet radio, the information in this book is not exhaustive, rather its purpose is as a guide for new users. It is not intended that it should be a standard reference manual and some liberties have been taken with the deeper meaning of some of the more technical aspects of 'packeteering'. Those readers interested in the technical side of packet radio are directed to study the various manuals available from such sources as the ARRL and the RSGB.

It must be stated here that both the authors and the publishers categorically deny any liability for any loss or damage to equipment or any consequential loss thereby. Most of the information herein is the result of personal experience, as well as advice passed on by others. We pass it on as truth because it was passed to us as truth. Since any technological hobby continues to develop, some of the information herein may be no longer applicable. This is naturally something over which we have no control.

We further advise that any software which you use, unless written by or for yourself, is either public domain or purchased by you. We cannot, and do not, condone the piracy of software, especially for commercial gain.

Our role in the production of this book has been one of assimilation of the technical knowledge required to ensure success in the first steps in packet radio, rather than to contribute any major original work. We would therefore like to duly acknowledge the many disparate sources who have contributed to our knowledge, and thus to this book. We apologise for being unable to name them all, but thank them none the less.

We have used the names of equipment and of manufacturers as they are used in everyday conversation between radio amateurs, and we acknowledge that many of them are registered trademarks.

If there was to be such a thing as a dedication, it should go to those selfless folks who give users a service, be it a mailbox, a node or a simple digi. To those we offer our heartfelt thanks.

Dave Coomber, G8UYZ
Martyn Croft, G8NZU
1989, 1990

Chapter 1

Introduction

"I'll play with it first and tell you what it is later"

It annoyed him, it really did. Just about every time now when he settled down to an evening in the shack, there he was. Sitting on the local net frequency as though he owned it. Spouting forth about the latest rig this, the standard of operating that, shouldn't allow Class B, should be morse only. And on it went. Oh, for the decorum and etiquette of a civilised conversation.

Why, he wondered, did this chap spend so much time on two metres if there was much wrong with it? Why didn't he take his kilowatt linear, his 17-ele Yagi, and his microprocessor-controlled rig and devote his attentions to some other band? Especially when some people get by on three watts and a vertical. Microprocessors? That was another topic for derision.

"Wouldn't have one in the shack" was a comment often heard.

"I'll play with it first and tell you what it is later."

This manual is intended as a helpful guide for the newcomer to get started in packet radio, or 'AX25' as it is often known. The authors have contributed something of their knowledge and experience, whether technical or plain operating, combined with the good advice we have had from the many helpful operators we have worked. The equipment quoted is that from our own experiences, and we hope that the contents will help steer the newcomer around some of the problems found when getting started in this fascinating field of amateur communications.

Packet radio arose, like many things, from several sources. The Rand Corporation of America conducted studies of packet networks in 1964. It arrived here in the UK in 1965 to be worked on by the NPL (National Physical Laboratory). But it was 1969 before any serious work was done by ARPA (Advanced Research Project Agency) in the USA. The first major radio experiments were conducted in 1970. A requirement by students on the Hawaiian Islands for data to be transferred without using the noise-prone (or non-existent) telephone lines led to the development of a network called ALOHANET. This featured computer terminals linked to a central computer by radio, with repeaters for those outstations (or network nodes) too far away for good direct contact. Here in the UK, the X25 protocol structure which had been originated by AT&T (American Telephone & Telegraph) was further developed by BT (British Telecom), among others, with additional input by the Open Systems Interconnection organisation.

A lot of early work was done on early data links by the Vancouver group of Canada in 1978-9. At that time, the mechanics of transmitting the data, for example the modem (which generates tones to transmit data over audio circuits), was not a direct part of the protocol specification, and the packet assembler/disassembler unit was separate. Even the exact nature of the packet was not quite as it is now, either. However, the notion was of interest to a widening group of interested amateurs.

In October 1981, the first ARRL Networking Conference took place. In the following month, six people with a passing interest met in Tucson, Arizona. The result of this meeting was the TAPR (Tucson Amateur Packet Radio) Corporation with the avowed aim of producing a complete packet terminal node controller (TNC), complete with modem and PSU.

The result was the TAPR TNC and the appellation 'AX25', which stands for

"The early TNCs were more primitive than those used today."

'Amateur X25'. The early TNCs were more primitive than those used today. Some required a modem to be connected externally, as well as having an external power source. However, the modern TNC has reached the point where the chip-count is getting lower and, particularly with the advent of the single-chip modem, it is now possible to have a TNC on a half-length card operating within the case of a standard PC; and even the TNC in your pocket. The following for AX25, the amateur version of the professional X25 data transmission protocol, has grown quickly. In 1983 there were a few hundreds of American users. A year later, a few thousand. A year after that, 10,000 users; and it came to the UK. After a slow start, it is now very popular with thousands of amateurs incorporating packet TNCs in their stations, and commercial developments have meant an increase in the facilities offered by the new units.

And so it continues . . .

Chapter 2

OK, what is packet radio?

"I'm sorry, Dave. I can't do that"

He switched off the rig and powered up his computer. Half an hour had gone by when the phone rang:

"Not on two tonight, then?", said a familiar voice.

"Nope; too much QRM, if you know what I mean."

"Yeah, he is a bit wide... about two megs wide I'd guess, from the way he's splattering up here!"

Over the years they'd spent many hours discussing a common interest in electronics, in amateur radio. And now computers. A quiet frequency on two metres had been the ideal. Sure, there was the odd QRM, some of it intended, but soon there would be another problem. Distance. A change in QTH would completely relegate their discussions to the telephone.

"What do you know about packet?"

"Not much. I hear it's very popular in the States but the equipment you need seems very expensive – what exactly does it do?"

"It seems to be a bit like RTTY, but more sophisticated. You need a controller: a TNC. It's like a terminal unit. That connects to your rig, and a computer can act as a display unit. You type your messages on the keyboard, the controller puts it all together, and the rig transmits it."

"Hang on, isn't this a bit like the ideas we've discussed before? If you were to put some sort of address, a callsign, say, at the start of the message, and the callsign of who sent it at the end, you could transmit the whole thing. But only the station it was addressed to would need to decipher it, everyone else could ignore it."

"Sort of like a network on the air, then? You should be able to do electronic mail. And file transfers. Perhaps conferencing. Maybe we should each get a TNC and try it!"

Packet radio is a means of sending messages in the form of digital data from one amateur to another. Actually, it will do rather more than that, but in essence, that's what it is. You could say that it is a glorified RTTY, but faster and with (usually) less visible errors. One vital difference is that more than one contact can take place on one frequency at one time. It is based on the X25 data transmission system. But where X25 is primarily used to send data on a point-to-point link (like a telephone line), AX25 goes a bit further and allows better addressing in the packet.

The data (usually text) is assembled for transmission in the form of 'packets' of a given length. These packets each have some sort of header (to start it off), an address (so the system knows for whom the packet is destined) and any routing information (so it knows where to go), some sort of control or ident (so it knows what it is), and some sort of tail (so the system knows that the packet is at an end). (For more information on the construction of a packet frame, please refer to chapter 7.)

Upon receipt of the packet, the receiving station sends a complex acknowledgment signal and, assuming reception of the packet was error-free, the next packet is sent. No acknowledgment, and the packet is sent again. And again, until receipt is acknowledged.

In fact, it is this perseverance and error-resilience that marks it out as being one very good way of sending data. No more 'RPT RPT' messages to the sender who has just dropped out of ionospheric range. Packet radio is a well-defined way of persuading a computer to do most of the work for you. The etiquette and method of the conversations between the computers is defined in the AX25 protocol and implemented in a device called a 'terminal node controller' (TNC). This contains some sort of microprocessor, as well as a program in memory, and a modem. If you have seen or used RTTY, then think of a TNC as an advanced form of terminal unit (like an ST5), and you won't be too far wrong.

It is this box of tricks which is the interface between your data and your radio. It will ensure that, regardless of whether you assemble your message on a terminal or computer, the data is of a uniform type that can be read by different machines. The protocol governing these digital conversations is defined in a series of layers. Each layer covers different aspects of the link, from the physical transmission to the user interface. And for the moment, that's about all the technicalities needed.

Given that computer shall speak unto computer and a need for maximum spectrum efficiency, a whole world of communication is possible. Mail, in the form of messages, can be sent via satellites to stations all over the world – and you don't always have to be in the shack for the reply, either. There are 'bulletin boards', just like you get on the 'phone line, where you can leave messages for one, or all, to see. Mail can be posted to, and collected from, amateurs in other parts of the country with systems which pass on the mail in the small hours of the night – while you get on and chase that rare DX, or whatever. You can, of course, use it to hold a 'conversation' with another station in the same way as RTTY or AMTOR. But the keen contest operator should not worry; win, lose or draw, there is no way it will replace speech.

The beauty of packet is its ability to keep trying in the face of stiff opposition from atmospherics and other channel users. Furthermore, if you have ever spent an evening of quiet frustration listening for 'other members of the club' or a particular friend, despite no end of traffic in which you have no interest, then look no further! With a few instructions to the TNC it is possible to compile a list of your friends from whom packets will be accepted, and then screen-out unwanted traffic. It is the ultimate filter.

As a common pool of knowledge, it is excellent. A message requesting help on some esoteric subject is very likely to strike a chord with some station somewhere. Pretty soon, a reply will put you right on whatever has been causing you grief.

In the same way as the mobile station can extend its range through a repeater, all packet radio stations can set their equipment to operate in a similar manner. Hopping from station to station, the packet can be retransmitted, with no loss of content, to its final destination. As you might guess, this is known as 'digipeating', from DIGItal-rePEATING. More recent developments allow a more efficient method of networking, and you may well hear such phrases as 'NETROM', 'TheNET', and so on. Whilst they all have their own individual ways, they may be fairly (if simply), regarded as an advanced form of digipeater.

As you may gather, it is very possible that the nuts and bolts of the whole thing may grab your attention or enthusiasm, and provoke a multitude of ideas and experiments in this exciting mode. If so, then this book has served its purpose. We do not consider it the task of this manual to detail all the highly technical details of this absorbing facet of amateur radio; researching any further information you may require is up to you.

You do not need a degree in computer science to operate a packet station, but a basic working knowledge of the transmissions is a great benefit. A rough understanding of baud rates and RS232-type serial connections is a help, but by no means essential, especially at the start. However, a little enthusiasm will go a long way.

Computers and packet

You might be wondering why, in a book about packet radio, we are talking about computers. Packet radio has a lot to do with computers. In fact without them, it is unlikely that packet radio could exist. But don't worry if computers aren't your 'thing'; you do not need to be, or become, a computer expert to enjoy packet radio. It is simply that an appreciation of some of the computer fundamentals will help to explain the mystery of packet as well as help you understand how or why something is not doing quite what you expected. These notes set out to give a few essential facts about the computer. They are not exhaustively accurate and some computer-literate types may consider them over-simplistic to the point of inaccuracy, but we have tried to express concepts that are, to some, an insurmountable barrier to the understanding of computers.

A computer, in essence, has three basic ingredients; a central processing unit (CPU), some memory (the ubiquitous ROM and RAM), and some sort of input and output device (nicknamed 'I/O'): see Fig 1. All this is joined to a common set of connections called a 'bus'. The CPU is at the centre of the whole thing, taking information from the memory, doing something with it, or reacting to it as an instruction, and putting the result back in the memory.

Memory is of two sorts: read/write, commonly called 'random access memory' (RAM), and read-only (ROM). Random access memory is what is known as a 'volatile' storage medium, that is to say that data stored in RAM will be lost when the power goes off, although a lot of the critical data can be retained by using a small battery to maintain the data in the RAM. Actually, RAM comes in two 'flavours', but this can be ignored for the bulk of this application. However, ROM data tends to be more permanent: in some ROMs, the data is put in at manufacture, while in other varieties the data can be changed electrically or optically.

The data accessed by the CPU may be regarded as being of two types: either plain old data which the CPU will manipulate in accordance with some programmed function, or instructions, upon receipt of which the CPU does whatever the instruction tells it. Since the CPU needs a set of instructions to tell it how to perform even the simplest of tasks, a 'program' of these instructions is stored in memory. In a TNC, the program (software) is held in ROM so that as soon as it is switched on,

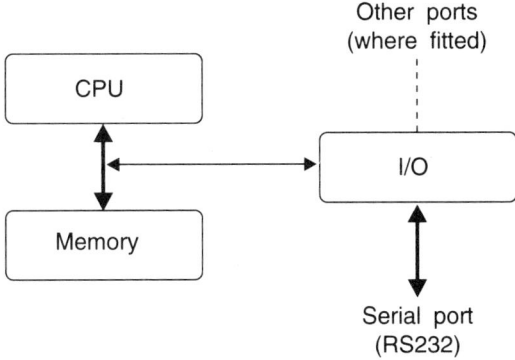

Fig 1. Block diagram of a computer

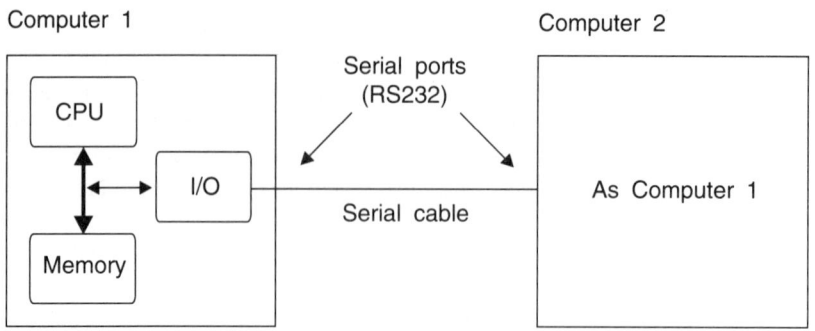

Fig 2. Transmission and reception of data via a serial port

its internal CPU knows what to do. The computer is thus essentially a manipulative device and it should come as no surprise that the CPU can also manipulate data to be sent to, or retrieved from, the outside world via the I/O (in/out port).

In the case of a typical home computer, the I/O facilities offered would included the ability to send parallel data to a printer, also to some mass-storage device such as a disk or tape unit. They would also include the transmission and reception of data via a serial port. It is often this port which connects to a great deal of complex equipment – like another computer. See Fig 2.

In fact, if two computers were just a reasonable distance apart, you could connect them together with a simple cable between each serial port. (This very trick is useful as a guide to whether that bargain-basement terminal from the local junk stall is working.) As you may have surmised, in this application of computers, they are too far apart and you have to use a TNC (or a modem if you want to spend money on using the 'phone). Serial data links deserve a little more attention, but first we have to understand the data.

Transmitted (or received) data consists of two states: 'mark' and 'space', which correspond to whether a voltage at a particular point is on or off, 1 or 0. In fact all the data, whether it is roaring around your computer or waltzing up a serial connection, is made up of a sequence of on/off states. This makes things simple for the computer, and difficult for us. The computer only has to deal with the ones and zeros: 'binary' is the proper term.

Because binary is the *lingua franca* of computers, all the information we want to manipulate and ultimately transmit has to be coded somehow. For this, computers use the well-known standard ASCII code. The code includes all the letters in the alphabet and the numbers, as well as some special codes necessary for control functions, and assigns each character a number in the range 0 to 127, although there is an extended set that takes it to 255. Thus the letter 'A' is represented by the number 65. In binary notation the number 65 comes out as 1000001. To be honest, you will see that hexadecimal (h or hex), is used much more than binary as the data highways (busses) are parallel, and it is more convenient for the whole character can be considered at one time. Decimal 65 is represented by 41h.

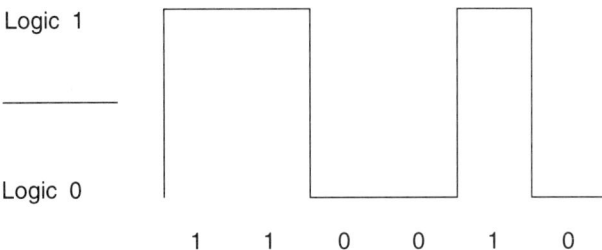

Fig 3. Binary equivalent of a figure 2

That's not as bad as it looks, because however difficult it may be to manipulate, it is surprisingly easy to transmit. Just send as a series of on's and off's, just like CW. Or even as tones, a high tone for a one, a low tone for a zero. A transmitted character might look a bit like that seen in Fig 3. For the computer's benefit, in binary, this would be 110010. In decimal that's 50. For our benefit the ASCII character represented is a number '2', or in hex, 32.

The TNC has more than a computer inside it. If you have ever heard packet data on the air, you will have noticed that it may have an almost musical quality about it, especially at the slower speeds. This is because the data is transmitted as musical tones, one each for 'mark' and 'space', which are the two elements which make up the individual character being transmitted. So the TNC must be able to generate these tones on command from its computer as well as being able to decode the data from tones received by the receiver. The device which accomplishes this task is a 'modem' and it is similar to those used on telephone lines for the same purpose. In a TNC there are different tones for different requirements. At VHF the tones of the American Bell 202 standard are used while at the more pedestrian HF, the standard is Bell 103. The fact that these particular tones are chosen matters not a lot, except that all TNCs understand them and can turn the tone sequences into data that means something to the computer.

The TNC thus acts as a 'bridge' between your transmitter/receiver and your computer/terminal. In many cases the shack computer acts as little more than a dumb terminal, and the TNC with its own CPU and modem is usually a little black box rather more adjacent to the radio than the computer. In the next chapter we'll discuss whether or not you even need a computer in the shack, and how to connect it all together.

Chapter 3

Getting started

"You gotta do the best you can with what you got"

He hung up the phone and pondered. The rig was mostly idle now. Computers still had many avenues for him to explore, but maybe here was a way that the two could be combined. But what did you do with packet? The same question had arisen over amateur radio. He'd seen the expensive transceivers, the enormous aerials and the RAE papers. Eventually he had been convinced. A licence to transmit wasn't the end in itself, merely a passport to a whole new world, a lifetime hobby. It would endure. Perhaps he should get a TNC...

It was some time before either did. A dip in prices and the move of QTH conspired to precipitate the purchase. He still didn't know why he'd bought one, and wished he hadn't when it came to connecting it up. Because it was a relatively new branch of amateur radio, newer even than satellite working, there were few articles written about it. Even fewer books.

But that was the challenge; wasn't it? ...

"You gotta do the best you can with what you got."

You will need three items to establish your packet radio station. A computer terminal of some sort; either a simple VDU or a computer that behaves like one. A TNC. And a radio. Oh, and some suitable cables to connect them all together. The basic layout is shown in Fig 4.

The radio

The choice of radio may deserve a little more thought than was first envisaged. A lot of amateurs have purchased surplus equipment such as Storno or Pye/Philips Communications transceivers, whilst several others use a portable or mobile transceiver as a base station rig. Surplus equipment represents a sound investment and is often available already converted to the amateur bands, if you fight shy of doing the job yourself. (Not all of us have the necessary patience or test gear.) For a modest price you can get 10 or 20W of FM signals on either the 144MHz or 432MHz bands. There are also some specialist radios which will handle the wide bandwidth necessary for 9600 baud transmission.

However, you should not take this to imply that it is impossible to use SSB as a mode of transmission although, because of the nature of the packet signal (as with RTTY), you will need some sort of tuning aid. Whilst there has been some activity on SSB, there seems little 'mainstream' activity, so you could be looking a long time for a contact unless you make a sked with some similarly-minded individual. If your radio pursuits are more HF than V/UHF or, more correctly, SSB rather than FM, the only marked difference with regards to packet is the speed of data transmission. On

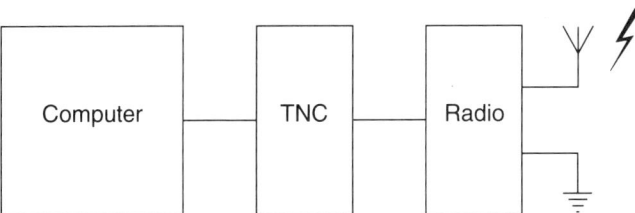

Fig 4. Block diagram of a packet radio station

HF it is a convention to use 300 baud; on VHF, 1200 baud is the norm. VHF and UHF systems tend to use FM in preference to SSB.

The big thing to realise is that it is not necessary to have a couple of hundred watts into an enormous beam aerial. This is because there are a lot of stations out there, and a lot of stations (network nodes) which will act as a means of relaying your signals onwards to the final destination. It is important to understand a little of the nature of the radio you are going to use, as there are a few 'minor adjustments' (mostly to do with the time it takes the radio to switch from receive to transmit and so on) which will ensure the success of your venture. These and other settings are dealt with later on.

The TNC

Which TNC you choose will depend a lot upon what you want to do with it and just which terminal or computer you may have, whether there is one for sale on your club's notice board, or even upon the selling skills of your local shop. Suffice to say that most TNC equipment advertised, whether in kit form or ready made, will do the job. The recommendations and experiences of the amateur fraternity can be used to great advantage here. It is not the intention of the authors to recommend any particular type of TNC. Those types mentioned (mostly based upon the TAPR TNC-2 and its clones) are quoted because those are the models we use, and upon which our experience was gained. It is perfectly possible that either or both of us would have got a lot further had we had different kit, but we didn't, so we can only quote from our experiences. Furthermore, the speed of development of electronic bits is such that any item we mention is likely to be obsolescent in short order anyway.

That takes care of some of the equipment. The rest requires that you have some bit of computer-related gadgetry that either is, or can be cajoled into behaving like, a computer terminal.

The terminal

Many people, like G8UYZ, have taken their first steps in packet using an old terminal unit of the sort more often seen on a mainframe computer installation. Others, like G8NZU, use the shack computer which pretends to be a 'dumb' terminal by running a terminal emulation program. In either case the terminal had to be coerced into talking at a rate of 1200 baud via the serial port (although faster speeds are in common use). This was accomplished, as is usual, with as many wires connected as the plug possesses and much cursing of burnt fingers and carpet. As you will find later, there are better ways of doing it. But it worked and that was all that mattered. It quickly got the station G8UYZ on the air.

Getting started 17

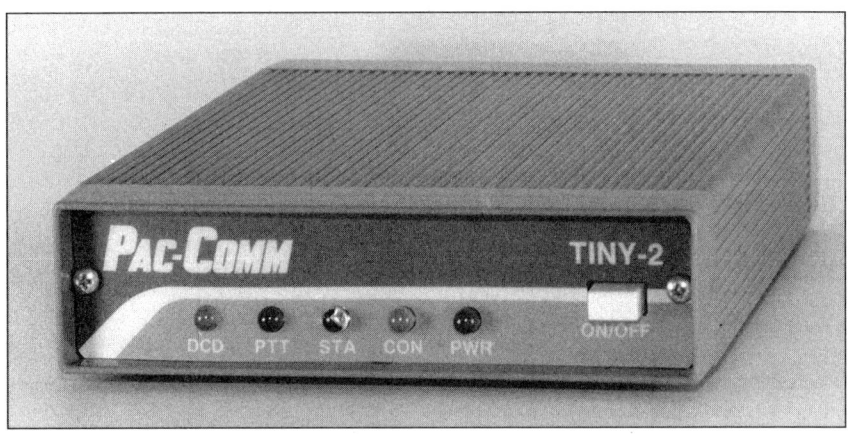

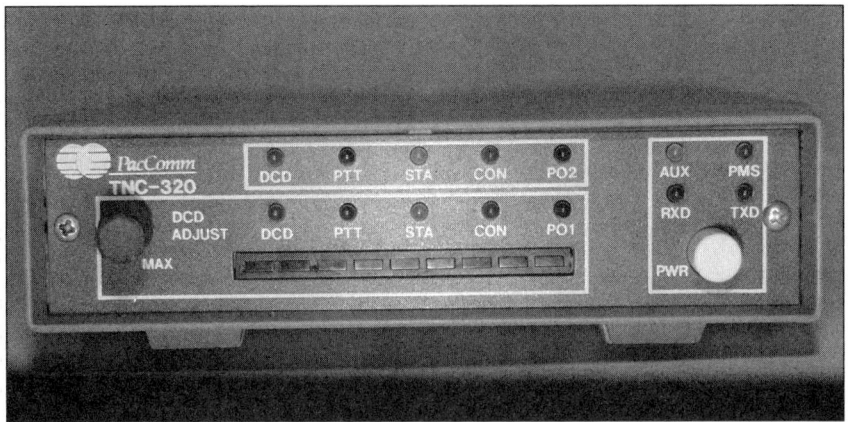

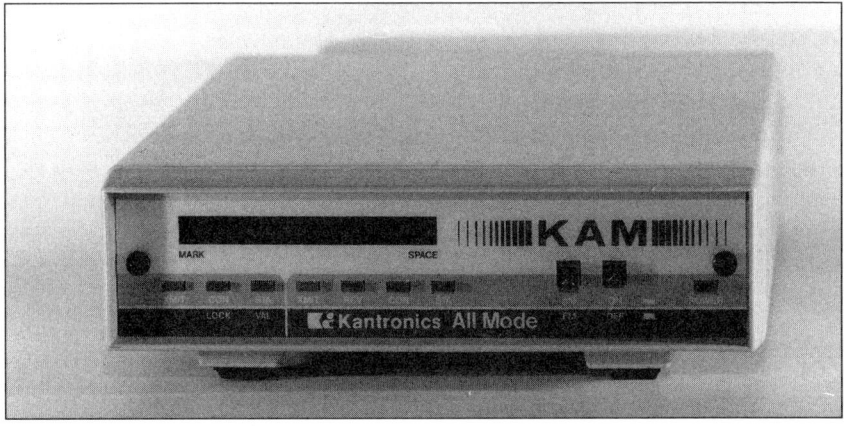

Three typical TNCs

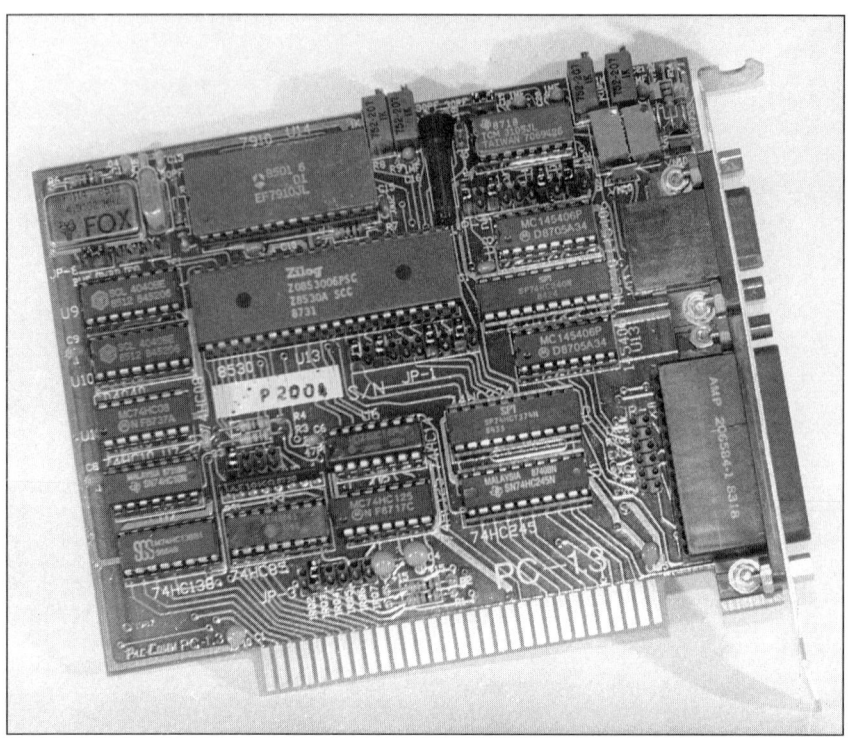

A TNC on a half-length PC card

Computer requirements

As noted above, G8UYZ started on packet with a computer terminal whose huge dimensions dwarfed the desk upon which it sat. The screen was the usual 80 columns wide but only 20 lines deep (most computer screens comprise 24 or more lines). But its real claim to use was a superb quality display; absolutely pin-sharp and crystal-clear lettering on the screen.

Because it was just a terminal, the only port on the back was an RS232 serial communications connector, which was used to connect it to the TNC. And, since this was a dedicated unit, it required no software. Using these old computer terminals is no great problem in packet, and since many large computer installations are regularly involved in the more or less continual upgrading process these days, it is fairly easy to pick up obsolete terminals at reasonable prices on the surplus market. (One was even found in a rubbish skip and it actually worked.)

The disadvantage of a terminal is its inability to transfer files and/or data. For example, you may want to read your mail and prepare your replies whilst disconnected from your local 'mailbox' (bulletin board system). However, as a cheap way of getting on air and holding a QSO, they have much to commend them. A more flexible solution is to use the computer already to be found in many shacks. Most

A miniature TNC using AX25 – note the size!

computers have software available to enable them to emulate a terminal, and many have a complement of advanced facilities. The more comprehensive the computer and its software, the better your chances of access to the files and software available on many mailboxes which often cater for the more popular machines. A second-hand PC is also worth considering; they need not cost a lot but considerations as to which sort is best are beyond the scope of this book. Try and talk to your local shop; he may have one that was traded in.

Whichever computer you use, it should, above all, have a serial port. Most PCs have one, often two (although occasionally you will see some symbol, like a mouse, for one of them). Some computers may require some form of interface or adapter card (for example, for early PCs, Sanyo MBC-55x, Apple II, Commodore 64, Dragon 32/64 and most Tandy/Radio Shack models to name but a few; talk to your dealer).

The serial port should be capable of supporting at least three connections:

Name	Description
TXD	Serial data FROM your computer TO the TNC.
RXD	Serial data TO the computer FROM the TNC.
GND	Signal earth, or common line.

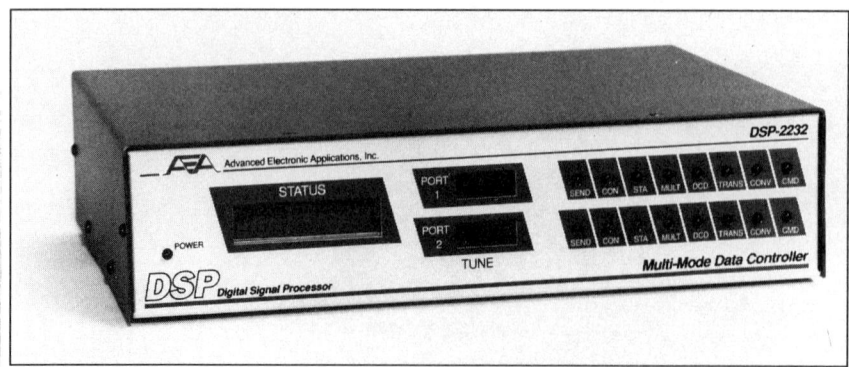

A multimode data controller, using digital signal processing (DSP), which can be set to simulate different types of modems

Since most of the packets of data we hope to receive are in excess of 80 characters long, your computer should be capable of displaying a generous size of screen. While, say, a portable computer with a small display will work, it can get a bit trying to read, and 80 columns is fairly common to many computers and terminals. Although many home computers were designed for use with a portable TV as a display, an Einstein or BBC, for instance, displaying 80 columns tends to look better than 40. With a proper monitor it is also a lot easier to read, and on packet you're going to do more reading than listening. Colour is also a help, but by no means essential.

Software requirements

'Terminal emulation software', as it is called, for your computer will ultimately depend a lot upon the computer you use. By way of illustration, there were two programs for the Einstein, lots even for the CP/M user, the BBC has a couple, and much software is available for the Atari and the Amiga. As with all software, the general rule is: the more popular the computer, the more software is available. A whole clutch of software is available for the PC and its compatibles. Much of it is in the public domain or available as shareware, which naturally cuts down the cost by a very large margin. The software requirements for the transmission of ASCII text over a packet link are fairly basic, but the ability to configure the parameters of the serial port (and thus what the computer says to the TNC) are useful.

For example, you may want to choose between baud rates of 300, 1200, 4800 and even 9600; SEVEN or EIGHT data bits; EVEN or NO parity; and ONE, TWO or NO stop bits. Some software requires that the handshake between computer and TNC be done in hardware (more on this later), whilst other packages are happy to conclude their business with the computer equivalent of a nod and a wink. The ability to transfer files, whilst not being very important initially, may well be a useful consideration at a later stage and this may require some thought as to the nature of

Getting started 21

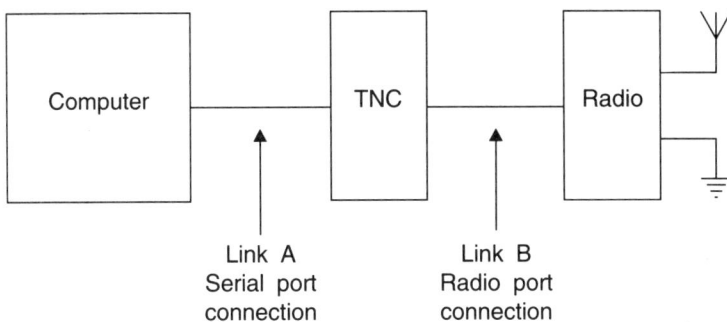

- **Fig 5. Block diagram of packet radio station showing the serial port and radio port connections**

the connections made in the cable. Many bulletin boards (or 'mailboxes' as they are also known) make available public-domain software for the BBC computer, PC-compatibles and others. They are also a useful source for updating your files on, for instance, modifications to your rig. Finally, those operators afflicted with either colour-blindness or a latent artistic tendency may find the ability to set up the colours used on the display to be of some advantage.

Data links

See Fig 5 (which is a version of Fig 4). Let's consider the Link A first. This is often called either the 'RS232 connection' or the 'serial port connection'. The terminal (or computer) communicates with the TNC via Link A at a speed no slower than the TNC sends data for transmission by the radio (slower speeds will work, but can have strange results depending upon the differences and the nature of the operator's typing skills). Many VHF operators set up Link A at a baud rate of 4800 and some at 9600 for a transmitted baud rate of 1200. The instruction manuals for your TNC will often detail both the ideal and minimum requirements for this connection. For a terminal or simple computer connection, you will need three or five wires; these connect TXD, RXD, signal ground, and maybe the 'handshaking signals' CTS and RTS as well. A lot will depend upon the exact configuration of your terminal/computer or, for that matter, the software and the purpose you have in mind.

Most TNCs are equipped with an industry-standard D-type connector, which has pins for all the connections mentioned above plus a few more besides. Of the usual 25 connections, many, you will be pleased to learn, can be ignored. Again, it depends upon the type of TNC. Instead of 25-way connectors, several of them now feature a 9-way D-type connector.

At the terminal/computer end, the variations continue. The BBC and the Einstein computer both have a 5-pin DIN (domino) connector whilst the modern PC uses both 9-pin and 25-pin D-types depending upon its configuration. In the next chapter we shall learn just how to establish this link but, as an illustration, when connecting G8UYZ's Einstein, all that was required was to short CTS and RTS together (thus forcing the handshaking) and run the remaining three wires (TXD, RXD and Earth) to the TNC. (The software was a file called TTY.COM.)

If we consider the TNC to be a 'black box' for the moment and continue to follow the data from the terminal on its journey to the ether, you'll be glad to hear that now a miracle occurs and audio tones and switching signals appear from the TNC. Link B is used to carry the audio tone to and from your radio and the drive to the PTT line. If you are a 'computer person', you might prefer to think of the TNC as being just like a telephone modem but connected to your radio. This is a bit oversimplified since a lot of 'thinking' (data manipulation) goes on inside a TNC.

Actually more goes on inside the TNC than was thought possible a few years ago, but, in practice, all TNCs do the same job. All the data in a received packet needs to be converted from audio tones (part of the modem's function) to digital data which is then read and interpreted, acknowledged if necessary (the packet might not be addressed to your station), converted to a suitable type and sent to the terminal screen. In fact, when just 'listening' (or watching) the traffic on the packet frequency, you can, if you wish, read packets addressed to any station. A cheerful and often instructive time can be spent just watching the traffic between mailboxes.

To send data back to other packet stations and communicate, the TNC needs control of your radio. It needs to listen to the traffic on the frequency (so the volume will probably need turning up a bit) and to be able to activate the press-to-talk switch. Trust it, it knows what it's doing. If it is unsuccessful, it will stop automatically (time out).

Chapter 4

Setting up the station

"Murphy was an optimist"

The phone rang. He could almost smell the flux smoking off the tip of the soldering iron.

"Is it 2 to 3, and 3 to 2 in a null modem cable?"

"Good question, I'll have a look. By the way, how long does it take one of these rigs to lock up in transmit?"

Power supplies, communications programs and cables had all been installed or crafted. Seldom was the shack in such a mess but the time had come for a test. Test, now there was something. How often had he really tested that ready-built rig? Transmits? – yes. Receives? – yes. On frequency? – yes. But test? Where was that dummy load?

"Where was that dummy load?"

When G8UYZ was getting interested in packet, G4OHK loaned a BBC computer with CommStar ROM-based terminal emulation software in his packet station. In this installation, the serial data link from the computer to the TNC had just four wires connected. On first thought you might think that it only needed two wires. The reasons for this apparent lack of simplicity are quite easy to grasp. It's all to do with 'handshaking'. You know, the electronic equivalent of "Hello – are you there?" and/ or "Are you ready?"

Some programs demand that the handshaking is done on wires separate from those upon which the data being sent. Others use the data lines to send "Are you ready?" messages. Each of these lines is graced with such titles as 'TXD', 'RXD', 'DSR' and 'DCD'. If they are not already, these names will become familiar friends during the next few pages. Look at a book on RS232 and its use (for example, a book on the use of a modem). It should tell you much about serial communications. You will also find our reference on the subject in appendix 3.

The terminal to the TNC

Establishing any serial data link, including the one from your terminal to the TNC, is as much an art as it is a science. Even the professionals sometimes get it wrong. However, take heart that for some as yet unfathomed reason, serial interface circuitry appears to be unreasonably resilient. But, be warned, there are no guarantees. Quite often, the only way to establish the link, and especially if you have not got the manuals, is to sit patiently with a soldering iron and try a few intuitive guesses. Start with the basic three-wire layout, TXD, RXD and GROUND. If you've got 25-pin D-types at both ends, then Ground is pin 7 and that is usually a safe bet. TXD and RXD are pins 2 and 3, and here is the first problem.

There are two things recognised by RS232C (or EIA-232-D as it should more properly be known). Generally speaking, a device is either a 'data terminating equipment' or a 'data communications equipment' (DTE or DCE). Depending on what the terminal thinks it's playing at, it may send its data on pin 2, or it may send it on pin 3; so in your lead, try 7 to 7, 2 to 3, and 3 to 2. If that does not work, try it the other way round and work up from there. Hopefully the worst you will see is a

26   Packet Radio Primer

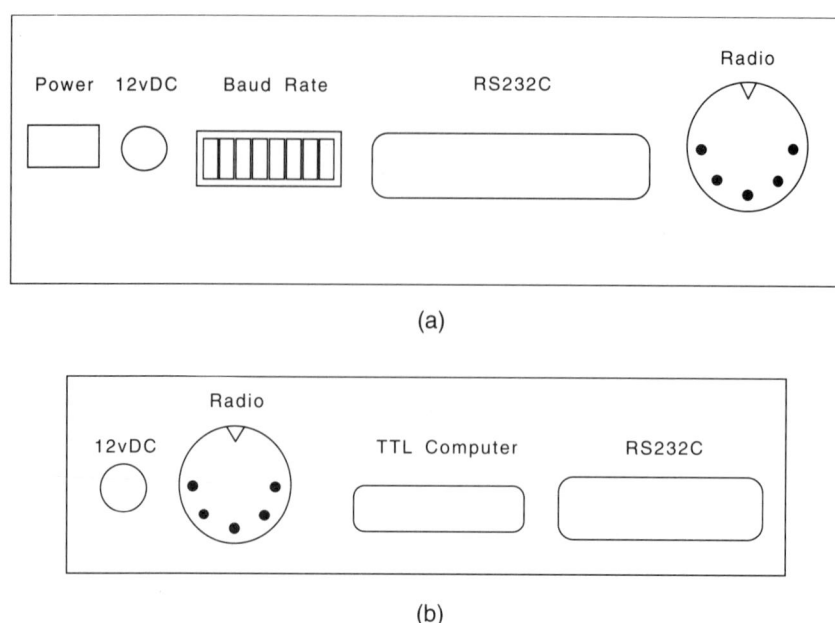

Fig 6. Typical rear panels: (a) TNC200 and (b) Tiny-2

blank screen. You might find it advantageous to have a stock of 8-core screened cable for these trials, although you just might get away with a commercial lead. And a lack of 'intuition' can be eased considerably by the use of a 'line monitor'. This is a little box with two D-type connectors and a row of LEDs which show just which of the lines are doing what. They don't cost a lot and are available from a wide variety of sources. A good alternative is a logic probe, or even an oscilloscope. Sometimes a multimeter can be used!

Some software (particularly that for transferring files) may need more than the minimum number of connections, and the mode of the TNC's operation may also dictate your connections. This is because some methods of handshaking employ what is know as 'software handshaking' which sends code characters in the data stream (usually Control-S (^S) and Control-Q (^Q)). This may often be seen as using XON and XOFF control codes. These codes are 'filtered out' and acted upon as necessary. Where a 'hardware handshake' is used, it can be more difficult to wire up, but is often more efficient.

This involves connecting at least the other two handshaking lines mentioned earlier: CTS and RTS. There are another two lines you should also know about: DCD and DSR. To give them their full titles, these are Clear To Send, Request To Send, Data Carrier Detect and Data Set Ready. Some computers just won't play if they don't get a signal on these lines. If you get this problem you may like to try the layout shown in Table 1 as a starting point. It links the handshake pins at the computer end together to fool the computer into thinking that all is well.

Table 1. Starting-point connections

Computer 25-way D-type	TNC
7 -------------------------------- Ground	
2 -------------------------------- RXD	
3 -------------------------------- TXD	
4, 5	(Short together)
6, 8, 20	(Short together)

Of course, you will have to rely on the software to accomplish the handshake (XON, XOFF); it might, depending on the baud rate, even manage without. Otherwise the computer, or the TNC for that matter, will have no way of saying "Whoa, that's enough for time being, thank you" and the result, as you might expect, will be garbled data. But at least you will be able to see something happening and can improve things from there.

An example connection

If you are handy with a soldering iron, making up the cable (Link A) can be a job which is definitely not difficult. As an example, a PC-type computer has the following lines available on at the serial port (it is a 25-way D-type connector) as shown in Table 2.

On the popular Tiny-2 TNC, a similar set of signals appear on a 9-pin D-type as shown in Table 3.

So to connect a PC-type computer with a 25-pin DB connector to the 9-pin DB connector on the Tiny-2 we will require a cable wired as in Table 4.

Programs like YAPP require a hardware handshake. Accordingly, several TNC makers have seen fit to make the wiring easy (possibly by modifying the internal

Table 2. PC computer serial connections

Pin	Mnemonic	Name
1	FG	Frame Ground
2	TXD	Transmitted Data
3	RXD	Received Data
4	RTS	Request To Send
5	CTS	Clear To Send
6	DSR	Data Set Ready
7	SG	Signal Ground
8	DCD	Data Carrier Detect
20	DTR	Data Terminal Ready
22	RI	Ring Indicator

Table 3. Tiny-2 TNC connections

Pin	Mnemonic	Name	TNC
1	DCD	Data Carrier Detect	O/P
2	RXD	Receive Data	O/P
3	TXD	Transmit Data	I/P
5	SG	Signal Ground	
6	DSR	Data Set Ready	O/P
7	RTS	Request To Send	I/P
8	CTS	Clear To Send	O/P
9	RI	Ring Indicator	(used for RF DCD)

Note: O/P means output from TNC; I/P means input to TNC.

Table 4. IBM PC to Tiny-2 (or TNC 200/220) connections

```
     D-25 (female)            D-9 (female)
     PC                       Tiny-2

      1  --------------------- 5*
      2  --------------------- 3
      3  --------------------- 2
      4  --------------------- 7
      5  --------------------- 8
      6  --------------------- 6
      7  --------------------- 5*
      8  --------------------- 1
     22  --------------------- 9
```

*Note that pin 5, Signal and Frame Ground, are the same in this instance. You could use the screen of the cable for this connection.

connections), so that a simple cable can be made (or purchased). For example, G8UYZ's old TNC200 has a 25-pin D-type with the following connections:

```
      1       Frame Ground
      2       TXD
      3       RXD
      5       CTS
      6       DSR
      7       Signal Earth (Gnd)
      8       DCD
      9       +12V Test   [a couple of milliamps only!]
     10       –12V Test   [           ditto              ]
     20       DTR
```

As long as you don't actually connect anything other than a test meter to pins 9 and 10, all is well. The connecting lead could not be simpler – 1:1 (ie pin 1 to pin 1 at the other end, pin 2 to pin 2, and so on throughout the cable (a seven-way screened cable is ideal). Leads are thus easy to make and not very expensive to buy. It might save you a lot of trouble.

The TNC to the radio

In cases where circumstances warrant extra screening, it is suggested that you use an eight-way screened cable and connect the screen at only one end. This prevents radiation from, and interference to, the serial link.

Link B, the radio port connection, can be made with a little less of a problem. In fact, having successfully negotiated the pitfalls of the serial data link and gotten thus far, this bit is a doddle. The TNC needs to be able to 'listen' and 'talk' to your radio. To do this it needs control of the press-to-talk switch, and connections to the audio in and audio out of your rig. For most amateurs we shall assume that we are on safe ground here. A quick delve into the appropriate manual should elicit a suitable socket for the connection of the signals in one fell swoop, and a trip to the shop should furnish a connecting plug. Failing that, you could try using the extension speaker socket and microphone connection: see Fig 7. A switch box to tidy it up would make a safe little project for the winter nights.

Taking the details of the Tiny-2 and similar TNCs again, the interface signals are available on a 5-pin DIN connector, shown in Table 5.

Thus to connect, say, the Yaesu FT290R to a Tiny-2, we will need a cable making up as shown in Table 6.

The audio connections, ie microphone and loudspeaker connections, might benefit by being individually screened from each other and the screens connected

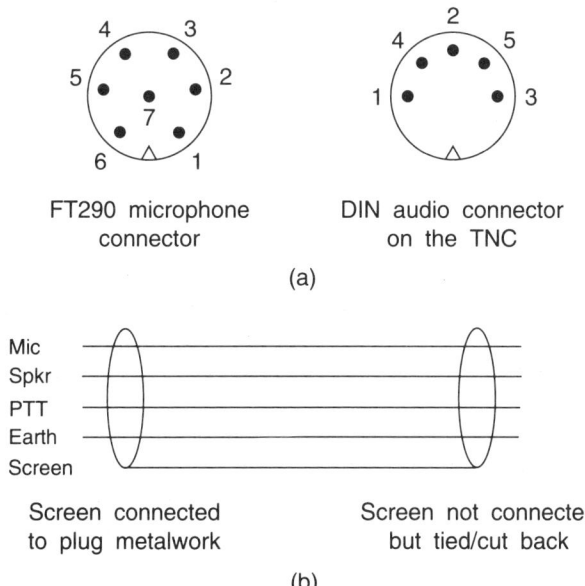

Fig 7. Connections between FT290 and TNC

Packet Radio Primer

Table 5. Tiny-2 radio interface connections

Pin	Name
1	Microphone Audio (from the TNC to mic i/p)
2	Signal Earth (screen)
3	PTT drive (see note below)
4	Loudspeaker o/p (RX out to the TNC)
5	Squelch i/p (not vital)

Note: you may have to take an audio feed from the loudspeaker socket. Be careful to get a good level of signal to the TNC, but do not deafen yourself! You could take a feed from the 'hot' end of the volume control if your radio does not have an audio output (but do be careful; you may invalidate any guarantee, so talk to your dealer). The PTT drive is often a transistor 'pulling' the PTT line towards earth, so read your radio's manual with care to ensure that this is what is required.

Table 6. Cable connections: FT290 to Tiny-2

FT290	Tiny-2
1 gnd	2
2 mic	1
3 ptt	3
5 spk	4
4 5V	NOT CONNECTED!
6 up	NOT CONNECTED!
7 dn	NOT CONNECTED!

to GND/Signal Earth. The connections should not be too long, ie as short as practicable.

Interference

A great many computers can develop a lot of radiated interference. Some computer makers screen their circuit boards and many PC-type computers are in metal cases which naturally cuts down the level of radiation. To minimise the noise, screened leads should be used for all connections. Common practice is to connect the screen at one end only (to avoid earth loops). In really difficult cases radiated interference may be mitigated by placing the computer in some sort of earthed metal framework, which will act as a Faraday cage. Some attention should be given to the actual layout of the installation; in particular, the routing of the cables and the positioning of the main units. For instance, it would be folly to place the computer too near a sensitive VHF preamplifier.

It may be possible to screen the interior of the computer, although this is not recommended without a good knowledge of the interior workings or the approval of the maker. For example, a small tin shield over the disc controller in an Einstein solved one specific problem.

Bonding the screen of the microphone connection to the case of the plug will also reduce interference in the sensitive audio circuitry. This screen should not be connected to anything at the other end. Naturally, the common (earthy) lead should be connected. On the data side, it might also be found useful to have access to the chassis (Frame Earth connection). This has been found to be very helpful in suppressing the ingress of RF interference. Using a screened connecting lead and making the screen the Frame Earth is a worthwhile enhancement.

Setting up the TNC

Many TNCs have the settings for both the serial port (Link A) and the radio port (Link B) made as switches. Figs 8 and 9 show the speeds for both. The big thing is that you should NOT set these switches whilst the TNC is switched on, although little damage is done setting the radio speed. It is vital that the two should be set so that the computer or terminal has time to display the data, hence the serial port speed should be equal to or faster than the radio port speed.

Many operators set the radio speed to 1200 baud (for VHF/FM) and the serial port speed to 2400 baud or faster. This is good practice, and should be adopted until you have the experience and confidence to conduct further experiments.

The switch-on

Now that the station is connected up, the temptation to go on the air is often irresistible, but there is one more hurdle. You need to verify that the data links are working correctly. Do NOT connect the radio yet. On powering up the equipment you should at least see a 'sign-on message' from the TNC. But suppose that all you see is a load of meaningless gibberish on your screen, or worse still, nothing at all. Assuming that the connections are correct, you may need to set up certain parameters governing the operation of your computer/terminal serial port so that it matches the operation of the TNC's serial port. These parameters cover baud rate, number of data bits, parity and the number of stop bits. There is a good chance that your TNC will have the following as its default setting:

Baud rate 1200, data bits 7, parity even, stop bits 2 (actually, one works as well).

You should set up the computer or terminal to give 1200 baud, 7 data bits, even parity, 1 or 2 stop bits and verify (or perhaps set, if you have some switches to twiddle) that the TNC radio baud rate and number of data bits is the same. In practice, 7 data bits, even parity and 8 data bits, no parity are the most common combinations, and a mismatch will often at least show that something is indeed happening. The only vital thing is that the serial link (Link A) between the computer and the TNC is set to the same speed, although this can be faster than the radio transmission rate.

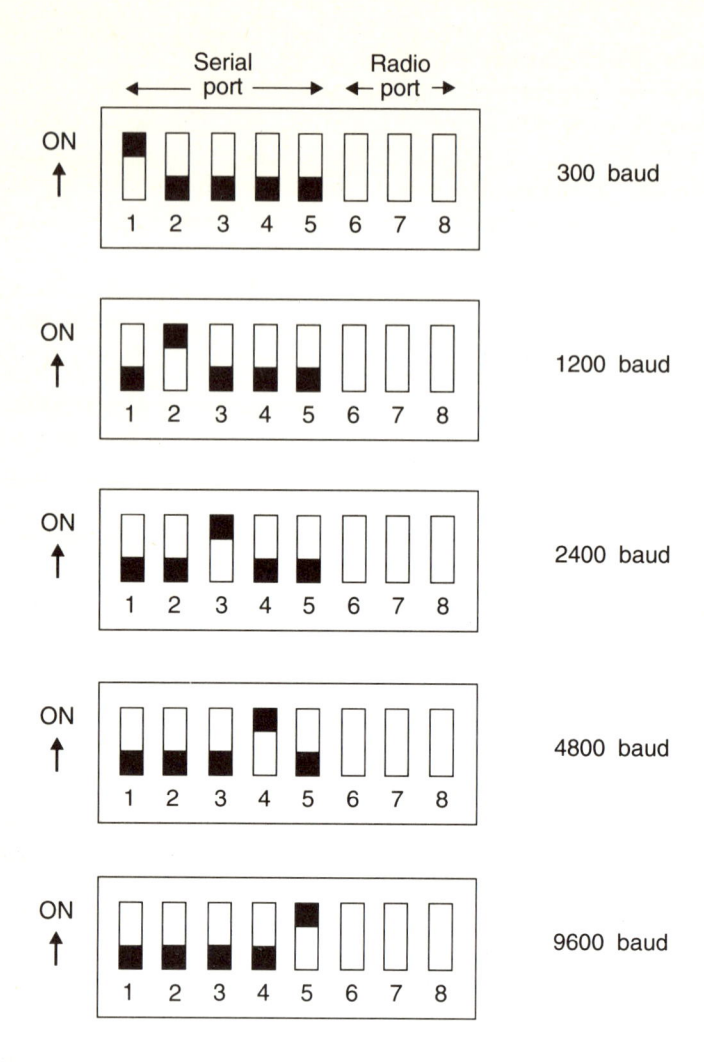

Fig 8. Setting the serial port speed

If you see what looks like a load of meaningless nonsense, adjust the computer (or software), settings for, say, 8 bits or odd parity. Eventually you will see plain text on your screen. This takes the form of the messages giving details of the type of TNC and various other information.

Cutting the branch upon which you sit

Beware! Do *not* be tempted to play with the TNC command settings governing the TNC's serial link parameters 'just to see what they do'. You could easily wind up with a chronic inability to make your computer talk to the TNC at all. The only way

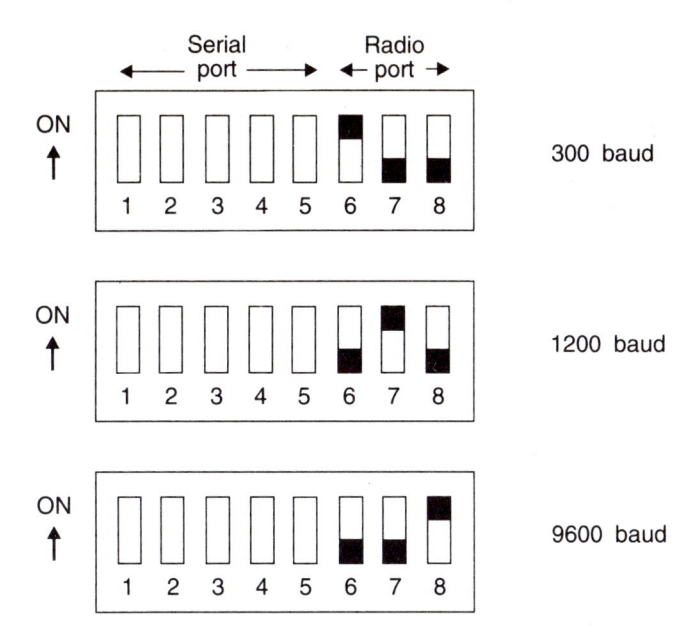

Fig 9. Setting the radio port speed

out, if you do make this mistake, is to switch off the TNC, open the case and remove the battery (or open the supply link); wait about five minutes (go and get a cup of coffee or something), then reconnect. The TNC will reset to its default parameters. This has happened to several frustrated packeteers!

It is now time to turn our attention to a few settings you must make to your TNC, before we step completely into the world of packet radio.

Chapter 5

Getting on the air

"You bash the Balrog, I'll climb the tree"

The computer screen burst into life as the TNC announced itself. Then, having proclaimed its existence, it waited.

"Try a command", he told himself.

But which command? There were so many. Something to do with the serial port, that way it won't start transmitting who knows what. He looked down the endless list of TNC commands and selected one that looked fairly harmless. The TNC was struck dumb by his insensitivity.

Two days later, a whole weekend, and his understanding of packet radio had broadened considerably. The TNC now keenly actioned his every command. Flicking smartly to transmit, sending a trill of data into the air, and listening intently for the reply.

Local packet stations responded eagerly with news of their endeavours to explore this new mode. Data zipped from station to station, digitally repeating the message to the next node in the empirical network. Numerous experiments, and they could be called nothing else, craved his attention.

"You bash the Balrog, I'll climb the tree."

It is a *very good idea* to spend a good while just looking at the packet traffic on a channel before trying a session yourself; just sit back and watch it happen whilst enjoying a congratulatory glass or cup of something. For without a doubt, the one thing which will tell all and sundry that you are a newcomer to packet is to connect up the station and start sending packets willy-nilly.

The TNC parameters probably will not be set ideally for your station's configuration, and the callsign held in the TNC's battery-backed memory is almost definitely not yours. A new TNC will usually arrive with the parameter MYCALL set to NOCALL and a quick squirt of packet down the rig will result in

```
HELLO WORLD from NOCALL
```

on everyone's screen. This may mark you down in others' eyes as a bit too enthusiastic, if not a complete wally. So first off, set up the TNC.

Before you can send a command to the TNC, you will have to attract its attention. This is usually accomplished by sending it a <Control-C>, often written <Ctrl-C> or ^C. To send this, press and hold the CTRL key and then the C key. The reply on your screen will be a prompt:

```
cmd:
```

Once you have the TNC's attention, there is a whole wealth of commands for you to try out on it. Don't forget that the TNC has a computer inside it. This is in addition to the one you may be using to talk to it. The TNC's computer possesses some 'intelligence' of its own and, for instance, does not always require the whole of a command to be typed. Many of the commands which the TNC recognises are used to change the numerous parameters governing its operation.

To make life slightly easier, the TNC will say what the value of the parameter was prior to any changes you made, eg:

```
cmd:MYCALL G8UYZ
MYCALL was NOCALL                       [this is the reply]
cmd:                                 [and returns the prompt]
```

Note: CAPITALS are not always needed and many commands can be shortened to a few letters. If you see this:

    cmd:MMYY GG88UUYYZZ

you will need to set the ECHO command by typing:

    cmd:EECCHHOO OOFFFF
    ECHO was ON
    cmd:

Note that you only type `ECHO OFF`. It just looks double on the screen because the screen display is duplicated by the TNC. Some screens (or computer's software) may actually need this facility.

If what you see on the screen is not as plain and simple as that, don't worry. You can set up either the terminal or the TNC so that they understand each other. Check out your TNC manual and the reference guide in the appendix and try a few *simple* commands. *Heed the warning at the end of the last chapter.*

A command badly typed or not understood by the TNC will evoke a reply of `Eh?` A little brief on its part, you will agree, but none the less perfectly understandable. You should try again.

It is possible to obtain the current state of a particular parameter by just typing in the command with no change, eg:

    cmd:MYCALL
    MYCALL G8UYZ
    cmd:

You can have quite a reasonable conversation with your TNC without ever going on the air! The parameters you enter will remain in effect until changed. The internal battery in the TNC will maintain the parameters in memory. Any changes that you do make, however, will affect the way that your TNC operates, so in these early stages you might like to make a note of what each parameter was before the change. That way if you suddenly think "I wish I hadn't done that", you can change back to the factory settings without resorting to the big reset.

Incidentally, there are two commands which might be of use, at least in the early stages of your learning: RESTART, which will go back to the opening screen without upsetting the defaults, and RESET which, as its name suggests, resets to the default settings all the command variables (like MYCALL to NOCALL and CT to a blank).

A few commands need to be set prior to operating and these may be grouped as 'essential' and 'definitely useful'. There are two types of command: that which affects the way the TNC works during a contact and that which affects the mode or operation of the TNC. For example, <Ctrl-C> will switch the TNC into its command mode. This is an 'immediate command' (ie one executed immediately for the operation or mode control of the TNC, not control of the data it handles). We will deal with any others as and when needed.

Essential commands (to get you up and running)

Note that some commands may be different in your own TNC and you should carefully read its handbook. See also chapter 6 on version 1.1.6 ROM. The capital letters represent the usual minimum of command. You could type `MYCALL` ... but you get the same effect with `MY` ...

MYcall

As might be expected, this is what tells the TNC your station's callsign. Typing `MY g9aaa` will put the callsign G9AAA into the TNC's memory and thus identify all packets from your station. The licence requires that you transmit a CW ident, and version 1.1.6 (and later) of the TNC2 software has the ability to do just that. (It will also do a lot more besides, like run your own personal mailbox, which will be dealt with later.) If your TNC is equipped with this version, you will need to set it up. This is done with the following commands:

CWIDtext

Type in your callsign.

CWid

A setting of E 18 will transmit your callsign every 30 minutes. (An 'A' instead of an 'E' will send it *after* the set time.)

CWLEn

Transmission speed. 6 (default) = 20wpm.

Note: If you have an early TNC which does not have the ability to send the necessary ident, you will need to upgrade by building or buying an ident unit. It is sometimes possible to do this in software. Fuller details on these commands are given later.

Definitely useful commands

BT Beacon Text

Every so often the TNC will transmit your beacon text automatically. The text of the message (120 characters maximum), is held in the BT buffer (an internal memory area). For example:

```
cmd: BT G9AAA located in Yorkshire, IO82WJ
```

This will put the message:

```
G9AAA ... etc ... 82WJ
```

on to all listeners' screens at intervals defined in B every (or after) n. See below.

Here we encounter our first problem. In early versions of 1.1.6, it was best not to bother with your beacon text, as the CW buffer used part of the BT memory allocation, and you could wind up sending the full buffer as part of the CWid. This could cause a few problems so the rule with the early ROM should be "Short is beautiful". Later versions (after 13/03/89 and 14/03/89) corrected this and other minor bugs.

If you are using a different version, and want to use the BText, it has to be enabled with the unconnected Beacon command:

B Beacon

This requires data on how often and under what circumstances you want the BText sent. For example:

```
B every n
```

(where 'n' is a unit of time, often in fractions of a minute). This setting will transmit the BText message every so often, regardless of whatever activity is taking place.

```
B after n
```

This setting will cause transmission of the BText after a time interval 'n', following the detection of any packet. It will therefore not take place as often on a busy channel. This is a very friendly way to operate, as it does not clutter a busy frequency with wasted packets.

A typical setting in a town or city might be:

```
B after 90
```

which in a TNC will send the BText about 15 minutes after the last packet was heard. Note: it is unfriendly to put in a ^G (Bell) character in your Btext message. Whilst it does attract people's attention, the chaos of having lots of random G-bells going off in the night can be quite unnerving, especially for the non-radio members of the family. Please use the G-bell with great restraint.

UNproto

This defines the nature of the beacon. It may be set to either CQ or Beacon:

```
UN CQ
```

will set it to transmit a CQ header on an unconnected packet.

        UN B

will set it to transmit the Btext. It is possible to send both forms through a series of digipeater stations. When using version 1.1.6, leaving it on the 'CQ' (and thus awaiting a call) is probably the best setting. When you connect to a station, it is possible that you will receive a message giving the status of that station.

The sequence of commands involving 'connect texts' and so on is as follows:

CT (Connect Text)
This is another of the text buffers like BT above and is sent when a connection is established.

        cmd:CT Please send a couple of ^G bells. I may not be watching
        the screen.

or

        cmd:CT I am not in at the moment. Please try later.

Some TNCs and software, like v1.1.6, allow storage of traffic, so it may be possible to have a message left for you to read when you return later on.

CMSG (CMeSsaGe)
Setting this to ON will enable the transmission of the CText.

CONOK (CONnectOK)
If you are out or doing something else, setting CONOK to OFF will automatically send a BUSY message to whoever attempts a connect to you and not permit a valid connection. If ON, a connection will be allowed.

CONV (CONVerse)
This is an immediate command which switches the TNC from command mode to the normal conversation mode, which means that what you type gets transmitted. When a connect is made, it will happen automatically (but you can do things about that, if you want).

C (Connect)
This is the primary command telling the TNC to go out and find someone:

        cmd:CONNECT G8NZU

The TNC will then send a sequence of packets to the stated station (in this case, G8NZU). If he is not otherwise engaged (with his CONOK off), his screen will show a line like this:

```
** CONNECTED TO G8UYZ **
```

at which point G8UYZ will see the CText from G8NZU on his screen, and the contact can continue. If his CONOK is off, G8UYZ will see

```
** G8NZU Busy
```

or a similar message. It is possible, and quite likely, that G8UYZ may need to send his packets via another station whose DIGIpeat is set to ON. In this case, the transmitting station determines the route to be taken:

```
cmd:C G8NZU via G9AAA
```

This will cause the TNC to transmit a packet to G9AAA and request a digipeat to pass on the packet. More digipeating stations can be nominated, separated by commas. If you wanted to connect via a network node, you issue a connect request to the node and issue another connect instruction to the station you want (further details in chapter 7). There are several variations to this theme and the complexity and development of the whole network is continually undergoing change, so we will restrict ourselves to the simple things for the moment.

The length of a packet can be controlled. It may seem odd that we might want to send longer or shorter packets, but it is important, especially on the more crowded parts of the band. The length of a packet is controlled by the Paclen command:

```
cmd: P 128
```

which will define the data in the packet as 128 bytes. When you type a long string of sentences, every 128 bytes the TNC will send them to the transmitter. 128 bytes is the default, but the maximum value is 255 bytes, which might be used to transmit a longish file on a quiet channel one evening. Oddly enough, P 0 represents 256 bytes. Today's crowded channels mean that 80 bytes (about one line of text), is probably best. Actually, there is another parameter which may be set depending upon conditions.

If MAXframe is left at the default setting of 4, by the time you've sent four lines of text, you should have had an acknowledgement for one or more. If there is a lot of traffic, and you cannot QSY elsewhere, you should try to shorten these values, which might actually speed up your effective transmission rate. In fact MAXFrame 1 is probably the most efficient.

A few tweaks (those minor adjustments)
Appearance on screen

There are a number of commands which make things a bit clearer on your screen. When monitoring packets (with MON set to ON), it will be seen that the header, showing the stations involved, is in the same line:

```
    G9AAA G8UYZ : Hi Martyn, how are you today?
```

Setting the HeaderLine command to ON:

```
    cmd: HE On
```

will produce

```
    G8NZU > G8UYZ :
    Hi Martyn, how are you today?
```

You may find this style of output easier to read, since the message text is separated from the addressing information. It is possible to add all sorts of information to the display, although most of the stations seen so far do not bother much.

If you want to keep good logs (on a spare disk!), with all the details of the traffic, you must set up the clock and a few other things first. For example, you may wish to see just which repeating stations are responsible for supplying you with the signal. The appropriate command is MRpt (Monitor Repeater).

If MRpt is OFF (and HE OFF), a line might look like this:

```
    G8UYZ > G8NZU: Hi Martyn!
```

regardless of the route taken by the signals. However, if MRpt is ON, the line might look like this:

```
    G8UYZ > G8NZU, G7ABC*: Hi Martyn!
```

The '*' indicates the station actually being received. Of course, in a log file, it is necessary to set up the date and time. This is done with the DAytime command:

```
    DA yymmddhhmm
```

where 'yy' is the year, 'mm' the month, 'dd' the day, 'hh' the hour and 'mm' the minute. So to set the date/time for half past nine in the evening of 27 June 1989:

```
    cmd:DA 8906272130
```

One interesting thing here is that you can display the date in both European or USA ways. This uses the DAYUsa command, which by default is ON so the date is shown as 'mmddyy'. This will show the date above separated by slashes:

```
    06/27/89
```

Setting DAYU OFF will display the date in 'ddmmyy' form, separated by hyphens:

```
    27-06-89
```

The command for putting this lot on a packet display is MStamp. Setting MStamp ON will produce:

        G8UYZ > G8NZU [27-06-89 21:30]: Hi Martyn!

So a fully optioned monitoring of a received packet, with DAY, DAYU, HEaderline, Mrpt and MStamp all set to ON, might well look like this:

        G8UYZ > G8NZU, G7ABC* [27-06-89 21:30]:
        Hi Martyn!

Performance adjustments

So much for the appearance of the information on your screen. What about any performance adjustments? If your radio transmitter is not one of the more recent types, and uses relays to enter the transmit mode (or it's just a little slow), it is sometimes preferable to set up some slight delay to allow the radio to actually stabilise in the transmit mode. It may also be necessary to set up a slight delay before the packet is actually sent to the transmit strip.

The command `TXD (TX delay)` forces a slight delay (milliseconds) between the PTT line switching and the emission of the packet to the microphone socket. It also allows the receiving station a little time to open its squelch circuits.

For example, an FT221RD seems to work better with TXD set to 30 (at least with the TNC200), giving a delay of about 300ms after pulling the PTT line down, before the packet is actually sent to the microphone input. This slight delay allows relays to drop and phase-locked loops to settle prior to the actual sending of data. Obviously, it is of use to find out how long your radio takes to lock up, as too long a delay is probably a bit antisocial, while too short could mean no contacts.

Once you get going, a friendly packet station should be able to give you some helpful feedback on the status of your end of the link while you adjust the parameters. There are other adjustments which can be made but, as these rely on a certain amount of experience, we will leave these to others to explain.

Some basic do's and don'ts

DO finish a live transmission text with a symbol. (You don't need it when using a mailbox). Convention so far has it that a chevron, '>>' is used. It really is a waste of time putting your callsign and his at the end of an 'over'.

DO make sure you avoid transmission before seeing his '>>'. We realise it can be a long wait for the next part of the message (you might even 'time out' – get disconnected – because the packets are lost or too much traffic or both), but it can save great deal of confusion.

If your software has provision for typing your reply as you see his message to you, make sure that you do not send the thing until you know he has finished. It is

Getting on the air 45

a good idea to stick to shorter messages, at least if the traffic is heavy or your contact is via a digipeater station.

DO *remember to issue a SENDPAK character (normally <Enter>) every so often (like just before the end of a screen line).* This includes those with computers using a 40-column display. You should do it at or before the text reaches the right-hand edge of the screen. Whilst failure to do this will not affect what is sent, it can look very odd at the other end.

DO *observe the official guidelines for the use of the UK packet radio network shown in appendix 6.*

DON'T *bother with the ^G bell unless it is part of a reply, or requested by the other station.* There was a time before a bit of common sense prevailed and the practice dried up, when a lot of stations were putting a G-bell in with the Btext. On a reasonably quiet night, it was possible to go deaf with the row. It might seem like fun now, but the bells, the bells . . .

DON'T *attempt any of the tricky stuff like multi-connect and so on, until you feel a bit more confident about the operation of your station.* Some TNCs default to USERS=4 but this should be changed to USERS=1.

It can be very confusing, but totally plausible, to attempt a multiple-connect. But unless you possess a compartmentalised mind, it can lead to a lot of heartache, frustration, guilt and disappointment, to say nothing of the confusion you will cause your fellow amateurs.

DON'T *tie up your local mailbox for hours.* It can clog up the works, and other users can get a bit upset. If you really want to sit down for a long session, try doing it at a time which is not quite so busy. Above all, remember that there are other users about and, like other modes, they can monitor what you are doing. Try to predict what will happen before you do it. However, packet is in its early stages and deserves to be experimented with. Don't be afraid to try something new.

A first contact

Hopefully you will have been monitoring a busy local packet frequency. In the UK 144MHz band plan this is often 144.650 or 144.675MHz (on the 432MHz band it is 432.675), and will have seen packets or beacon text from a local station. Let us assume that in this case the local station is G8UYZ and that he is not busy at the moment. G8NZU attempts to connect. With the `cmd:` prompt on the screen, type `CONNECT G8UYZ` and press the <Return> or <Enter> key (in future we will assume that the <Enter> key is pressed). On the screen this looks like:

```
cmd: CONNECT G8UYZ
```

46   Packet Radio Primer

Note that it is not always necessary to type in the full word. A `C` will institute a connect.

At this point all manner of wonderful things happen as the TNC takes over. The TNC recognises the CONNECT command and assembles a packet of data to send to G8UYZ which will request a connect. The radio is switched to transmit and the packet transmitted. The TNC now listens for a returning packet from G8UYZ. If there is a lot of traffic on the channel, this may take some time. However, the TNC will not listen forever; it will eventually 'time out' if no response is heard.

With a link established, the message

    *** CONNECTED to G8UYZ

appears on the screen. G8UYZ will also be informed of any progress and see on his screen:

    *** CONNECTED to G8NZU

The TNCs in both stations have automatically switched to CONVERSE mode and a connection has been established. Whatever is typed at one station will appear, eventually, on the screen of the other, and thus the messages can be passed. The big question, of course, is how to stop it and return to the command mode.

Because the TNC has switched modes (to CONVERSE from command mode) it needs a ^C (<Ctrl-C>) sending to attract its attention again. This should result in the now-friendly `cmd:` prompt. At this point simply type `DISCONNECT` (to be honest, a simple `D` will suffice). The TNC will make the necessary excuses to extricate you from the conversation and respond with:

    *** DISCONNECTED

after which you will be back to the `cmd:` prompt.

Hopping around

So much for a simple, direct contact. Suppose, however, you were out of direct-contact range with the required station. Without a little help, you could have problems. For example, G8UYZ is not currently well placed to get anything without a little help, but the area is filled with helpful guys who leave their 'DIGI' on. DIGIpeating is a facility which allows a TNC to pass through the station without that station's help; it is automatic. A lot of operators will indicate their particular state (DIGI on) in their beacon text and your contact will be 'via' their station. For example, if G8UYZ wants to contact G6DKT, it will have to be via some other station, so he types:

    `cmd: C G6DKT via G4RVK`

Getting on the air 47

This method can take a little while, depending upon what G4RVK is doing at the time and the amount of traffic on the channel. The 'via' (which may be shortened to 'V'), will ensure that the TNC at G4RVK will not bother with an attempt at connection, but pass it on to G6DKT. If it is a tricky route, several digi stations may be required, so I would type:

    cmd: C G6DKT via G4RVK, G1OLB

There is, of course, a slight problem. There is a maximum of eight digi stations permitted and, on a busy night, you could wait a long time for a simple "Hello Mum" message to be passed, to say nothing of accumulated errors and/or timeouts on the path. It was once mooted as to the feasibility of doing Land's End to John O'Groats, but getting stations in the right place (at the top of a mountain or two) has so far proved too difficult.

It is at this point that a bit more technology comes to the rescue; it's called a 'node'. (For more details on nodes, see chapter 7.) The principle of connections is just the same. It's just that all the error correction is done by the node/s through which the message is passed. On a five-digi hop, improvements of 800% have been quoted. To use one you have to connect to your local node in the same way as you would a simple direct contact:

    cmd: C GB7AP

or whatever your local node is called.

This will get you a reply telling you that you have connected and it sits there awaiting your next command. Notice that it is a GB callsign, as the nodes may be the subject of different licences. You then type in your next connect as if you were driving *it*:

    Connect G6DKT

And thus you get connected to the station you wanted. Obviously, the time taken for a packet to be sent will depend upon the traffic density and thus it is still possible to time out, but at least you do it with fewer errors. At the time of writing, a lot of operators are addressing the problems of a complete network which does all the connections for you. When it does, a contact from one end of the country to the other will be like working the station just down the road, albeit a bit slower.

Chapter 6

The packet postman

"2 + 2 = 5, for moderately large values of 2"

The nearest packet mailbox system became a constant source of delight and information. Like its well-established cousins on the 'phone lines, the bulletin board was the place to post news and views for one, or all, to read. The mailbox was the gateway to a national, and even worldwide, network of other amateurs using packet radio.

With each new contact he reflected that the TNC is the epitome of the polite conversationalist. Listening attentively to the chitchat. Waiting demurely to interject its contribution. Apologising for talking at the same time as another. Patiently seeking an encouraging response. Diligently repeating its message should the recipient not hear.

"2 + 2 = 5, for moderately large values of 2"

Amateurs have been passing messages by word of mouth for many decades. The problems arising from this method of communication include the fact that we are not all blessed with perfect memories and odd notes taken of items of interest tend to be stacked up in odd corners of the shack. Similarly, a long chat over some quiet frequency might well turn on the modifications to or construction of some electronic gadget of mutual interest.

Wouldn't it be easier to pass that information in a manner that can be stored easily and cheaply and then printed out at need? It can take hours to describe the construction of something as simple as a power supply or aerial, but a file with the details can be passed via the computer with ease. Furthermore, the keen but solo programmer can face hours of frustrated work when that published listing does not work as it should. A file of corrections passed by another who has succeeded speeds his enjoyment by no small measure.

The local bulletin board (at the end of the 'phone line) is often used for just such similar purposes outlined above. You just dial in and leave a message, get help on a problem, maybe get some software and generally have a little harmless fun. However, it does cost you a 'phone bill every now and then. Wouldn't it be nice to do it on the radio? Well, you can!

The growth of various mailboxes (or 'bulletin boards', as they were often known) has meant a lot of activity on both the 144 and 432MHz bands. It is an ideal method of getting a quick message to the members of a group. A mailbox can be particularly handy as a club noticeboard, giving details of forthcoming events and messages to members such as 'hot' DX news.

The GB2RS news bulletins are forwarded automatically to all mailboxes in the UK every week, and may be viewed at leisure, rather than getting up early (!) on a Sunday to listen.

The mailbox

Many of the newer types of mailboxes are based upon two main systems, the WA7MBL and the W0RLI (although there are now several other types, such as the AA4RE, G4YFB and G1NNA). For those who have been involved with packet radio for some time, the commands required by a mailbox will seem logical, and will have been learnt in stages at each upgrade. For those who are new to packet the thing can look a bit daunting, to say the least!

For those who have not seen an mailbox from 'the other side', here is a brief description of the installation. The host computer is often an IBM PC-compatible using hard disk drives and a number of serial ports, each connected to a TNC and an amateur transceiver (typically one per band).

A mailbox often uses a multi-tasking file manager to allow multiple ports to be operational simultaneously, allowing for two or more users (on separate bands) to access the database files at the same time. This appears transparent to users, and there is no indication that another station may be connected at the same time as yourself, although a small message is often displayed if another station is trying to connect. Where you see this message, it is good form to quit as soon as possible and allow others to use the mailbox.

The mailbox usually offers the facility of:

(a) Local, national and international mail;
(b) national and European 'flood' bulletins (sent to all mailboxes);
(c) a central file server of both text (ASCII) and binary files.

In addition, there are a number of 'local' functions unique to each system.

It is important to remember that each mailbox can be configured to suit the sysop (system operator) and the users, and as such this guide refers only to the basic commands. You should consult your local sysop for information regarding any regional configuration. Should you encounter a menu which restricts your activities to personal mail services only, you should again contact your sysop for further details. This guide covers the facilities offered to those afforded full access. We have structured this section as 'looking over the shoulder' of an operator of the old standard 'MBL, with notes to help translate some of the shorthand jargon which inevitably creeps in when speed is more important than verbosity.

Be sure to decide on *one* 'home' mailbox and keep to it. Do not wander between different systems, as this will only serve to confuse the network, and maybe delay your mail. This is particularly important where you have declared a 'home' mailbox, as eventually it will be possible to send mail to a station without the necessity to declare the destination. Remember that the mail is almost the same on every mailbox, and if you want files from distant systems you can do this automatically by using the REQDIR and REQFIL commands which are covered later – use the network!

Never DX into a mailbox. All GBR bulletins are received by all mailboxes, so it really is not worth the effort.

For the purposes of this illustration, we will use the commands of the WA7MBL system. Other types will have similar commands, but often differ in the facilities offered to the user.

Having called your local mailbox you will be greeted with a log-on message which will read something like:

    [MBL-$] Hello, New User,

    Thank you for calling this <named> Mailbox serving the <named> region. To become a 'regular' user, please type N and then your name.

    Menu A,B,D,H,I,J,?,K,L,R,S,T,U,V,X

At this point you should input the 'N' (it stands for 'name'), and a prompt will appear at which you type your first name. This usually indicates that you intend to use this as your 'home' mailbox (although some boxes have a specific command for this), and you will be afforded an entry in the user database, which is updated each time you call. You are encouraged to use both upper- and lower-case characters when sending to the mailbox. Always wait until you see the menu or prompt before entering your next command, as it is easy to lose synchronisation and become confused because the mailbox is acting upon a different (earlier) command to the one just entered.

Example session on a mailbox

Accessing your local box is simplicity itself and, for the purposes of illustration, we will call up G8UYZ's local box, GB7MAX, which operates on both the 2m and 70cm bands. The original was done on 70cm, and is taken from the log of the contact. For clarity, the subject concerned a modem which did not appear to work.

    cmd: connect gb7max

After a while, if the mailbox is not busy, a reply will come back onto your screen:

    [MBL-$]
    MAXPAK Mailbox

    Dave You have unread mail.

    rm   *[This means 'read my mail'; up comes the reply:]*

    Msg # 21070 Type: P Stat: Y
    To: G8UYZ From G1DIL Date:17May/1713
    Subject: This and That...
    From G1DIL @ GB7MAX

    Hi Dave,
    Good news about the modem. Do you mean that your driver software

       will not support V21 and V23, or the host that you call on the
       phone. I thought ProComm handled most speeds - if not there are
       other Public Domain / Shareware program that will do the job.

       Best 73 de Andy.

And now for the reply:

       sp g1dil   *[Send (a) Personal message. The mailbox replies:]*

       Enter subject for Msg # 21260:
       this & that   *[So much for the title; now the content:]*
       Send Message. Use ^Z or /EX to end:
                                [ie to terminate, use /EX or Ctrl-Z]

       Hi Andy,
       Thanks for the Note.

       I think the problem is one of different speeds, ie., 1200/75,
       which Procomm does not seem to handle.

The rest of the message is not important so we will skip to the end.

       73, Dave, G8UYZ
       /ex

       cmd: *[Now sign off]*

       b
       *** DISCONNECTED

Note: personal messages cannot be read by all and sundry although, obviously, they can been seen when input and read by the sysop.

At this point, you may see a general message transmitted from the bulletin board. This is the sort of thing which can tell you what's going on. It shows a beacon message from G1DIL-2, and indicates that it was sent out to GB3AP for onward radiation:

       GB7MAX Status, GB7AP-1:
       MAXPAK Mailbox
       Mail for:
       G1DIL, G6LYE, G6CMV, G6VAT

You may see a slightly different version. This illustrates another general message from another mailbox and shows with the asterisk that it was received via G4OHK:

```
GB7nnn-2MAIL,G4OHK*, AP2:
MailboxTamworth
```

The W0RLI mailbox

The W0RLI mailbox is similar in operation to the 'MBL, but has a different command set, and is (or was) not noted for its ability to handle binary files (programs) easily. It does, however, feature a few other things that the 'MBL does not, and the debate as to the best will probably be going for a long time. The following sample showing the commands available is shown below and also illustrates a connection made via a local node:

```
cmd:c g8vpr-8                           [a local node]

***CONNECTED TO G8VPR-8

##CONNECTED TO NODE G8VPR-8(G8VPR) CHANNEL A

***WELCOME TO G8VPR, KA NODE***
ENTER COMMAND: B, C, J, N, or Help

c gb7sam-2                              [ie connect the mailbox required]

###LINK MADE                            [reply from the node]
[RLI-10.01-CH$]

21.43z. Greetings, Dave, welcome to the Stoke area mailbox,
located in Clayton, Newcastle, Staffs (IO82vx). SysOp is Dave
G3TJP.

Type H for Help.
NE toggles 'expert user' status

You last connected at 1924z on 890510.

Please use NH command to enter your home mailbox.
Example: NH GB7SAM

21.43z OK, Dave, Stoke Area Mbx :
(B,C,D,H,?,I,J,K,L,M,N,P,R,S,T,U,V,W)

    H                                   [get help data]
```

And so the contact continues. The sample shows that there was a text file called 'help.bat', which could be downloaded or viewed at will. It should be noted that the facilities offered by particular mailboxes may not be identical, as it will depend upon the requirements of the users at a particular location and the sysop, to say nothing of the software used.

Downloading files

To a large extent, this will depend upon the facilities of both the mailbox and your software and is thus difficult to illustrate. It also depends upon the willingness of your fellow users to upload something for you to download. See appendix 2 for command details.

A personal mailbox

Several of the available TNCs are, or can be, fitted with a system which makes it possible to leave a message for the recipient in the manner of a full-sized mailbox. It is a bit like an automatic telephone answering machine, or like a bulletin board. The commands are very similar in operation, if not in name. A sample log of a contact (to a KAM TNC) with a personal message system looks like this:

```
cmd:c g8vpr-2
***CONNECTED TO G8VPR-2
**WELCOME TO G8VPR'S PMS***

***I WILL RETURN YOUR CALL ASAP***

ENTER COMMAND:
B,J,K#,KM,L,LM,R#,RM,S OR HELP

    H                                        [get help data]

    B                   Mailbox will disconnect
    J(heard)            CALLSIGNS with daystamp
    J S(hort)           HEARD CALLSIGNS ONLY
    J L(ong)            CALLSIGNS WITH DAYSTAMP AND VIAS
    L(ist)              LIST MESSAGES YOU CAN READ
    LM(ine)             LIST YOUR MESSAGES
    K(ill)#             DELETE MESSAGE NUMBER#
    KM                  DELETE YOUR MESSAGE/S
    R(ead) n            DISPLAY MESSAGE NUMBER n
    RM                  READ ALL YOUR MESSAGES
    S(end)(call)        SEND MESSAGE TO CALLSIGN

ENTER COMMAND:
B,J,K#,KM,L,LM,R#,RM,S OR HELP

S G8vpr                                  [send a message to him]
SUBJECT: testing testing

ENTER MESSAGE - END WITH CTRL-Z OR /EX ON A SINGLE LINE

Hi, Brian,
```

```
This is yet another test for the benefit of my log file, which
has been giving me some trouble of late.

73, Dave.
/ex
```

```
MESSAGE SAVED.
```
                                              [unless I kill it, he'll get it!]

Note: I had to try the above several times before getting it securely in the log for inclusion in this text; hence the message.

Many of the PM systems will behave in a similar manner.

It is important to realise that it is not possible to send messages to anyone but the operator of the PMS, and it might fairly be regarded as a sort of answering machine – *not* a mailbox substitute. However, it will be seen from the above that, in addition to the 'answering service', these systems may offer the newcomer a chance to practise a bit before going on a big board, even if it is only a case of leaving a message for the operator/sysop.

I happen to know that G8VPR's PMS is not version 1.1.6 software (it is a KAM), so the commands are not all the same to a user. Users seeing a v1.1.6 or later will thus encounter a different set of commands, although, it must be repeated, the functions are likely to be the same. Fortunately, the 'help' is just about constant. For example, someone getting into G8UYZ-2 will be met with the following:

```
Logged on to G8UYZ-2 PMS

CMD (K/K/R/D/B/H/?)
```

The options at this point are as follows:-

```
Bye         Disconnects the user from the PMS.
```

The user will get a message saying "logged off" and about 10 seconds later, the disconnect command will be issued.

```
Help / ?    This displays the commands.

Kill n      This will delete the number n message from the
            memory.
```

If successfully issued, you will see **Message Erased**. If you got the number wrong, you will see **Message not found**. If the message is not to or for the user's callsign, you will get **You cannot Kill this Message**.

```
List        Lists the 10 most recent messages in order:

Number, Month/Day, From, To, Subject
```

> Mine   Lists the 10 most recent messages to you.

Obviously, this will not be a lot of good unless the 3rdparty command is ON, which is should definitely NOT be (special licence!). Users should not even bother trying.

> Read n   This will cause the display of the message n.

If you try to read a message not for you, you will see You cannot read this message.

> Send xx   where xx is the callsign.

If 3rdparty is OFF, only the correct MYPcall (G8UYZ-2 in this case) will be allowed. Any other will cause the display of No third Party mail allowed. The maximum Subject length is 28 characters. Message length is limited by available storage. Text is terminated by either ^Z or /EX on a clear line. If successful, you will see Message Saved displayed.

Accessing another PMS

Another sample from the log shows a bit more of a personal message system in action:

```
cmd: c g1dil-2 v g4rvk                          [Notice the V]
***CONNECTED TO G1DIL'S PMS
CMD (K/K/R/D/B/H/?)

S                                               [not bothered about callsign]
Subject: Hi There
Message: Thanks for the help !!
73,
Dave
Please note my new PMS G8UYZ-2.
/ex

Message Saved                                   [confirmation of message]

CMD (K/K/R/D/B/H/?)

L       (list messages)

18      0204    g8uyz   g1dil hi there
17      2003    g6yak   g1dil Greetings
16      0131    g3szg   g1dil EI land
14319 bytes free
```

```
Next message number 19
CMD (K/L/R/S/B/H/?)

R 16                                    [try to read someone else's message]

YOU CANNOT READ THIS MESSAGE                           [the result!]
B                                                        [bye bye]

- Logged off
***DISCONNECTED
```

The following shows a good use of the CText message on his normal station as a 'missed me' message:

```
cmd:c g4rvk
***CONNECTED TO G4RVK

If no reply, please leave message on my PMS, G4RVK-2 - Dave.

cmd:D
***DISCONNECTED
```

Now to get into G4RVK-2:

```
cmd:c g4rvk-2
^G***CONNECTED TO G4RVK-2                       [notice the ^G bell]
s                                                         [Send]
SUBJECT:
Hi Dave
MESSAGE:

I now have a PMS v1.1.6. And this is to see if I can get mine to
work. Is your digi ON ?

73, Dave
/ex

MESSAGE SAVED                                           [I made it]
CMD (K/L/R/S/B/H/?)

B
- Logged Off

cmd:D
***DISCONNECTED
```

TNC ROM version 1.1.6

Versions after 1.1.6 of the PacCom TNC ROM feature a few of the essentials and a few more features which are of considerable use. As was mentioned in 'Setting Up', it is a requirement that all stations send an identification in CW at intervals of not less than 30 minutes. Pursuant to these requirements, the software now includes commands to do just that. They are as outlined previously, but included here for completeness.

CWid

This is your callsign to be transmitted. In some versions it took up part of the BText buffer, so it really was best not to use the BT function. (Why bother sending your callsign in both modes?) Now to determine when/how often it is to be transmitted:

CWIDTEXt

This is very like the BText, in that the text may be sent either 'every n' or 'after n', where n is the interval. This interval is in 100-second units. This must be set, as the default setting is 0, which switches OFF the function.

To send your CWIDtext out to a waiting world every half hour, set it thus:

```
CWID E 18
```

CWLEn

The speed of callsign transmission is set up by this length, where n is set to 6 by default. Speeds can be set between 40wpm (n = 3) and 10wpm (n = 7). Where n = 6, a speed of about 20wpm is set. Versions after 13/03/89 of the ROM are equipped with a few extra commands:

NOLogon on

This retains the PMS sign-on message, whilst

NOLogon off

will delete the PMS connect text.

STExt

This may be used to set the PMS sign-on message (in a similar manner to CText).

V

displays the firmware version and date.

ADrisp off

This strips the header information (who it is from and where) from the displayed, received text. Default is ON.

CRAfter off

will strip the carriage return whilst monitoring incoming frames. The default is ON.

Using the PMS as the sysop

In addition to the ident above, version 1.1.6 includes facilities to enable you to receive messages in your absence without leaving the computer on. To set one up takes a little patience and the following commands set in the TNC. If someone tries to connect as if you were in attendance on the system, he would call you in the usual way. However, if he wanted to leave a message, there has to be some way of showing that there *is* a difference between your usual callsign and your letter box. This call dictates the callsign to be used by callers wishing to leave a message. It is set by using the following:

MYPcall n

where 'n' is the callsign secondary station ident (SSID). For example, G8UYZ-2. You should *not* set it to be the same as MYCall. It can cause no end of confusion!

PMS on/off

If you turn it off, any messages left for you will not be lost. Default is OFF. If it is OFF, callers will see

```
*** G8UYZ-2 BUSY
*** DISCONNECTED
```

on their screens. Messages are stored numerically from 1 (this is set back to 1 on the issue of a RESET command).

3rdparty

This *must* be set **OFF**, since there are separate licences for operators of bulletin boards. Setting it to ON will allow your 'box' to act in the same way as a mailbox and this is specifically prohibited. As the owner of the MYPcall, you can send any message to anyone, but messages IN may only be left for *you*.

The other commands are:

Kill n

This will delete the number 'n' message from the memory. If successfully issued, you will see `Message Erased`. For example

```
KILLOld
```

kills the oldest 10 messages.

List

Lists the 10 most recent messages in order: number, month/day, from, to, subject.

Mine
Lists the 10 most recent messages to you.

Read n
This will cause the display of the message 'n'.

Send xx
where 'xx' is the callsign to whom it is sent. This could be handy for leaving a message to a specific caller. Of course, it helps to know *if* someone has dared to use your PMS. If a message is on your box, the STA light will flash slowly, which indicates a message waiting (at least it does at G8UYZ). Reading the message/s will reset the flasher, but not necessarily wipe them from the memory.

A few other useful commands

There are a couple of other commands which control the various delays and frames but, as this is a beginner's guide, it might be best to leave them for others to explain. Nine times out of 10, the default settings are all that is required. There is, however, a useful one which may help in the display of dates on the screen:

AMonth On/Off
If ON, the month in dates will be shown as 'Jan', 'Feb', etc. If OFF, they will be '01', '02' etc. This may be helpful in conjunction with the DAYUsa command: (ON = mm/dd/yy; OFF = dd/mm/yy).

A sample of using my own PMS

As G8UYZ can usually reach G4RVK without going VIA anyone, it might seem silly checking to see if he is allowing digipeating. However, it would be useful to connect to myself via him, so I can see what my own PMS shows up as, but first I set up my own box. And as you can see, it is fairly easy to play about with, and with a bit of thought you get the desired results. A subsequent CONNECT to myself by digipeating through a local station confirmed that it works as I want it to:

```
***CONNECTED TO G8UYZ-2

Logged on to G8UYZ's Personal Message System
CMD (K/L/R/S/B/H/?)

send g8uyz                              [send a message to myself]
Subject: test
Message: hi - is this working ?
Message saved:                          [seems to work]
B                                       [bye bye]
```

Now to read my mail. These commands are entered straight from the keyboard without connecting to anyone:

```
L         [list what's in the box]

2         00/00  G8UYZ  G8UYZ  Greetings        [test message]
3         00/00  G8UYZ  G8UYZ  test             [that's the one!]
14694 bytes free
Next Message is number 4
```

Now to read it, to check if it really is the one:

```
cmd:read 3
Posted : 00/00/00 00:00
From: : G8UYZ
To    : G8UYZ
Subject: test
hi - is this working ?
```

[the 'POSTED 0's' mean I have not set the clock/day/date]

```
cmd:send g1dil
Subject: test
Message:                                         [blank entry]

L                                                [List messages]
2         00/00  G8UYZ  G8UYZ  Greetings
3         00/00  G8UYZ  G8UYZ  test
4         00/00  G8UYZ  G1DIL  test
14666 bytes free
Next Message is number 5
cmd: kill 4                                      [delete message 4]
Message erased
cmd:k 5                                          [try shorthand]
?bad                                             [does not work]
cmd:kill 5
Message not found                                [there is NO msg 5]
```

Now I can put a message to all who access my box:

```
cmd:send ALL                    [general message to all users]
Subject: Welcome
Message:Welcome to my PMS
^G^G Please leave your message after the tone . . ^G Thank You.

/ex
Message saved                              [now to read it back]
```

```
r 5
Posted 89/03/03 20.53
From:    G8UYZ
To:      ALL
Subject: Welcome
Welcome to the G8UYZ Message board
[Beep]Please leave your message after the tone . .[Beep] Thank
You
/ex
cmd:
```

It is worth noting that any corrections you make when entering messages will also be seen when displayed. A <Backspace> will be displayed as a ^H, so it is worth trying to get it right. If you don't, delete it and do it again. Your users will thank you, and don't forget, a little humour can do much in showing a bit of individuality. Note: in this instance, I used the ^G bell.

Chapter 7

Packet protocols

"Don't knock paranoia, it's paranoia that keeps you on your toes"

He connected to the local node. Only a few miles away, its signal was strong. Via two more nodes, the packet of data sought its destination and elicited an acknowledge from the distant TNC. The monitor speaker in his shack echoed the warble of the modulated data; in between bursts he could hear the same old squawks on adjacent channels. He disconnected the speaker. It might annoy him, but it didn't trouble packet. The message on the screen was clear and uncorrupted:

```
*** CONNECTED G8UYZ>>G8NZU :  Hi Mart.
```

"Don't knock paranoia, it's paranoia that keeps you on your toes."

What's in a packet?

It is not the intention of this guide to detail exactly all that happens when you press the <Return> key. However, an appreciation of some of the complexities might not be out of place.

Data is sent in 'frames'. There are three main types, each with a different purpose and, if they were languages, they would have different dialects in that a particular type of frame may have different meanings depending upon just what is in the Control byte. All frames contain some commonality. They all feature a Start Flag, Address Data, Control Data, Check Sum and an End Flag. The Control field is a group of eight characters and contains something of interest to one end or the other. For example, the type of frame may be herein specified, or sequence data, or maybe an ACK (acknowledgement).

However, between the Control and Check Sum, there may be something to tell the network what level is in use as well as the information message you type in. Consider for a moment the fact that you have been monitoring the local net (Information (I) – frames by the score), and want to send a CQ.

Your TNC will issue a frame saying "here I am" – but not expecting any acknowledgement – so it sends a U frame. The U frame comes in different flavours, depending upon just what is expected of it. It is used to control and order the protocol when a connection is *not* yet made. Typical uses are Connect Request, Disconnect, CQ and beacon texts. In this case the Control field contains the same sort of thing that beacon messages are made of – UI.

As will be seen from Fig 10, there are an appreciable number of bits to be sent (up to 600 in some instances). The most significant contributor to the length of the frame is the Address field, whose length will depend upon such things as whether you are using a digi or two. The length may be made up of a TO address (seven bits), a FROM address (seven bits) and a VIA address (up to a maximum of eight seven-bit words). The seven bits are made up of a six-digit callsign and an SSID (secondary station identification).

When you send a packet of information (the messages you send out), an Information (I) frame is dispatched, and in this case, the length of the Data field will depend upon how many bits you have set in your PAClen command as well as the

68 Packet Radio Primer

Field type =	Flag	Address	Control	Check sum (CRC)	Flag
Size (bits) =	8	Less than 560	8	8	8
Information contained	01111110	Callsigns, secondary IDs of source and destination, incl digipeaters	Type of frame	Calculated for each packet	01111110

U frame and S frame

This is what is added to make an I (information) frame:

It goes in HERE

Field type =	PID Protocol ident	Information
Size (bits) =	8	Up to 256
Information contained	Network-specific data	This is what you type in your packet. It is set by PACLen command

Control field:

I frame: | ID 2 bits | 3 bits | 3 bits |

S frame: | ID 2 bits | 3 bits | 3 bits |

U frame: | ID 2 bits | 6 bit code |

Fig 10. A packet radio frame

address data. Contained in the CheckSum field is a 16-bit number. This number is calculated by the sending station. Having sent the packet, you have to get a reply; and here it can get complex. If the receiving station, which does the same calculation, does not get the same calculated sum as that sent, an error can be assumed and the 'received ack' is not sent; the receiving station dumps the frame and the transmission is repeated.

If all is well and the receiving station sends an immediate readable reply, you might get another I frame with some of the Control field set a particular way. Otherwise, you might only get the third type of frame, called the Supervisory (S) frame (see Fig 10). This, you will see, looks just like the U frame, except it has different data in the control field.

An S frame is used to signify to the transmitting station that the receiving station is or is not capable of receiving any more I frames for whatever reason. The S frame can also send an acknowledgement for up to four I frames (used, for example when sending a file or a really long message with short PAClen).

When you disconnect, your TNC will send another U frame and the process is terminated. There are several variations, mostly to do with the contents of the Control field and its effects. What has not been mentioned are the timers and unacknowledged frames. This is where a simple explanation can get complex, and that is not the aim of this book. Those with a need to get deeper into this fascinating world are advised to look for the manuals from the ARRL (via the RSGB), like *The AX25 Amateur Packet-Radio Link Layer Protocol, Version 2.0*, first published in 1984 by the ARRL, or even Tannenbaum on Networks.

There is a command in the TNC called TRACE. When Trace is set to ON, you can usually see exactly what is in a packet. What is important to note is the fact that for the purposes of keeping the data an even value, the hexadecimal values sent are double those actually required for display, so there are two columns of the stuff. On the left is 'as transmitted' and on the right the translation.

It can be most instructive, and a useful time can be had looking at the sheer volume of data being transmitted, as well as what it all entails. Try doing it by hand (or using a computer), and you will begin to appreciate the magic of packet.

A note on nodes

NET/ROM (or its clones like TheNet) is a networking firmware which will convert the ubiquitous TNC-2 (or clone) into a network node controller which in use acts rather like a digipeater. It is intended for wide-coverage digipeater sites, and not usually as a device for the end-user at home in the shack.

A node provides the user with a means of quickly being connected to other nodes, and thence to the desired station, rather than using the digipeater 'via GxXXX, GxYYY' command. It provides true 'store-and-forward' facilities (it remembers what you've told it and acts accordingly) over what is known as a 'transport level' (level 3 of the OSI model) link, which vests in it a certain amount of automation. It supports cross-frequency and cross-band working (like going in on 70cm and out on 2m) without the need for lots of hardware, so it is easy to install – even if this does mean another licence. It is faster than the usual multi-hop techniques used when connecting to a friend at some distant point using the VIA command on someone's digi. The reason for this increase in speed is due to the way in which it acknowledges a packet. A multi-hop digi signal is acknowledged by the destination station (end-to-end acknowledgement), which can mean that it is easy for a packet to get lost on a long haul on a busy night, whereas in a multi-nodal hop the acknowledgments are between nodes, which is much quicker and leads to less errors. See Fig 11.

A node can be fairly regarded as an intelligent repeater. For example, in addition to connecting you to your chosen station, it does a few housekeeping functions of

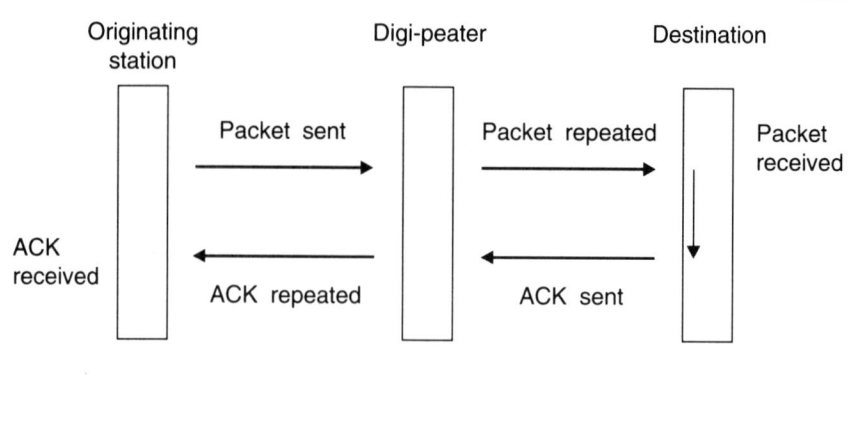

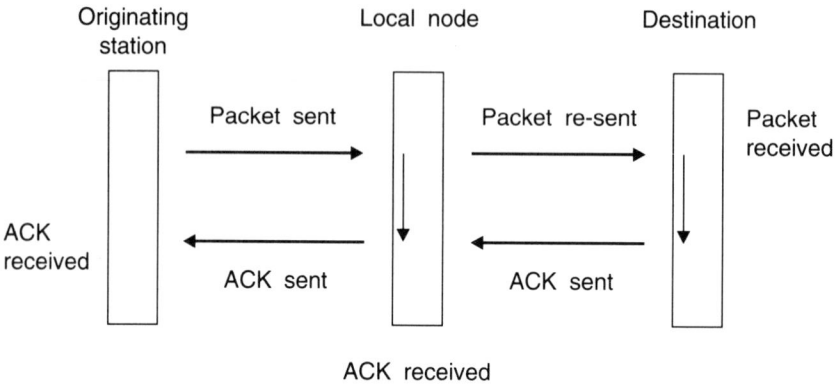

Fig 11. How an acknowledgement is handled by a digi and a node

its own. Every hour or so, nodes broadcast to one another, and thus a node listens to its neighbours. If it hears a new node (one that is not in its route table), it will automatically add the new node to the table and assign it a default setting as to its 'goodness' or quality of path. If it fails to hear it the next time round, the node will be eventually deleted from the route table. If it is again heard, the quality is again assessed and, if necessary, the route table is updated. So the chances of a path to some distant station are good as the information is rarely less than an hour old.

Commands

The firmware supports the following user commands: CONNECT, CQ, INFO, NODES, and USERS. For any of them, the entire command verb ('CONNECT') or just a fragment ('CONN' or 'CO' or 'C') is allowed. Any command parameters must be separated from the verb and each other by one or more spaces. The maximum command length is 80 characters. Commands must end with a carriage return.

The CONNECT command

The CONNECT command (C for short), is used to request a circuit to another node, a downlink to another user, or a connection to the node's host terminal. To request a circuit to another node, use the syntax:

    CONNECT node

where the 'node' must be the callsign or 'mnemonic identifier' (eg CAN72) of another node that is known to this node. (Use NODES command to obtain a list of all known node callsigns and identifiers.) For example:

    cmd:c can72
    *** CONNECTED to CAN72

To request a downlink to another user, use the syntax:

    CONNECT usercall <[VIA] digicall>

where 'usercall' is the callsign of the user station, and 'digicall' is the callsign (or alias) of a node or digipeater. Notice that you can use the 'VIA', but it would be more correct to connect to each node in turn. If digipeaters are used, the use of 'VIA' is optional ('VI' or 'V' are also acceptable). Digipeaters may be separated by either spaces or commas. For example:

    CONNECT PP72 via CAN72
    PP72:GB8NZU-3} Busy from G8NZU

In all cases a successful connection is announced by the message `Connected to callsign`. The message `Failure with callsign` indicates that the specified node or user did not respond after a number of attempts. `Busy from callsign` indicates that the node or user responded but refused the connection request. Other possible error messages are `Node busy`, `Circuit table full`, `Link table full`. These messages indicate a lack of resources in the node – the user should disconnect and try again later.

An in-process CONNECT command is immediately aborted if another command or a blank line is entered before the connection you have requested is established.

CQ command

As its name implies, it is possible to cause the node to issue a CQ call on your behalf. Replies will be received in the usual way. However, it is not universally regarded as good practice unless it is the only way to get out.

NODES command

The NODES command (N for short), is used to display the node's route table, which is the list of station nodes it knows it can hear.

To display a list of other known nodes listed in the route table with their mnemonic identifiers and callsigns, use NODES on its own:

```
cmd:nodes
CAN72:G8VPR-7> Nodes:
AP21:GB7AP-2      AP72:GB7AP-7       BM1:G7AXC-1      BM2:G7AXC-2
BM7:G7AXC-7       BNOR72:G6GUH-7     BRX:G4AKZ        CD21:GB7CD-2
DIL72:G1DIL-7     LRG21:G0GDR-2      LRG72:G0GDR-7    LX2:GB7LX-2
MALV21:G6CMV-2    MAXMailbox:GB7MAX  MLVN21:G4FPV-2
```

USERS command

The USERS command (U for short), displays a summary of who is using the node:

```
u
CAN72:G8VPR-7> TheNet Version 1.0 (290)
Uplink(G8UYZ)
```

The heading of the USERS display also shows other information, such as the version of the firmware in use at the node, and the amount of free RAM space (shown in parentheses, and expressed as a number of 32-byte buffer segments). It is worth noting that the suffix may not be just what you typed in; this is a suffix added by the machine, which will change it when needed (two identical callsigns would lead to total chaos).

After the heading, the USERS display may also show the active circuits and links, using the following formats:

Uplink (who is it from?)
Downlink (who is it to?)
Circuit (other nodes used)
Host (this node)

The '<—>' symbols seen in some versions represent active 'patchcords' in the node that connects uplinks, downlinks and circuits (and possibly the host terminal, if any). The lines which do not contain the '<—>' symbol represent users who are in command mode at the node. All this latter depends upon the actual firmware in use.

INFO command

This command (I for short) provides general information on the node. For example:

```
info
CAN72:G8VPR-7>
```

```
            Cannock, Staffordshire
            432.675 MHz 1200 baud
```

Here is another example:

```
            info
            RP72:GB7RP-7> Alport Height, Derbyshire. IO93FB
            432.675 MHz - 1200 Baud.
            * DANPAC * For info and membership contact Dennis G0KIU @ GB7DAD
```

To round off this short guide to nodes, here is a copy of my log getting into CAN72:

```
            cmd:c  can72
            *** CONNECTED to CAN72
```

``` n```	*[nodes command]*

```
            CAN72:G8VPR-7> Nodes:
            AP21:GB7AP-2    AP72:GB7AP-7    BHAM:G7BGP-1    BM1:G7AXC-1
            BM2:G7AXC-2     BM92:G7AXC-11   BNOR72:G6GUH-7  CAN22:G8VPR-2
            EDG21:G6JVS-2   EDG22:G6JVS-3   GLOS21:GB7GH-2  MAXBBS:GB7MAX
            RP72:GB7RP-7    RP91:G4KLX-1    VPQ90:G8VPQ-10  WP90:G0KNR-9
            WP91:G0KNR-1    WRGBBS:GB7WRG
```

``` r```	*[routes command]*

```
            CAN72:G8VPR-7> Routes:
              1 G8VPR-2 240 16
              0 GB7RP-7 10 1
              0 GB7MAX 10 1
              0 G6GUH-7 10 1
```

``` u```	*[users]*

```
            CAN72:G8VPR-7> TheNet Version 1.0 (249)
            Uplink(G8UYZ)
```

``` i```	*[info]*

```
            CAN72:G8VPR-7>
            Cannock, Staffordshire
            432.675 MHz   1200 baud
```

``` cmd:d```	*[disconnect]*

```
            cmd:*** DISCONNECTED
```

One of the uses to which a collection of nodes may be put is the fast dissemination of information on a particular subject. Take, for example, a group of DX enthusiasts who might well use PacketClusters.

DX PacketClusters

What are they? As many a DXer has been heard to say quite recently, "possibly the best aid to DX chasing since Marconi fired up his first rig".

DX PacketClusters are relatively new to UK amateur radio activity, but are one of the fastest-growing interests around. The basic idea of a cluster is to provide assistance to enable DX information to be disseminated to users of the system in near real time. To avail oneself of the facilities of a cluster it is necessary to log on to a system by doing a normal 'connect' as if you were connecting to any other station on packet radio. If you have a cluster in your local area, you can connect direct, but it is also possible to connect using routes through nodes on the packet radio network. One beauty of the system is that you only have to connect to your nearest cluster to obtain the facilities of all of them, because if the clusters are linked it has the same effect as if you are logged on to all of them simultaneously.

When you are connected you will get a welcoming message from the cluster and it will ask you to supply some personal details such as your name, QTH, latitude and longitude etc. Once these formalities are over you can get down to the serious side of the cluster.

Why not let a user give you his side of the story? . . .

DX clusters: a user perspective (from G4VXE)

"Anyone who has read the *Complete DXer*, written by W9KNI, will know that the art of DXing is all about maximising the odds of working 'that new one'. You can do this in many ways; having your station in tip-top form is one, being a superb operator is even better. But whatever, you need to be well-informed about what's 'going on' on the bands at any particular time. On your own you can listen to one or, if you're smart, maybe two bands at the same time, looking for that bit of elusive DX. However, the DX cluster system allows you to make the most of the DX information that is being fed into the network by many of the top operators around the UK.

"Is the system effective? Well, the simple answer is 'yes'. One evening I was tuning around one of my favourite bands, 18MHz. Suddenly on my screen a message appears:

```
DX from G4DYO: 18080.0 V47NXX
```

Very nice, I think. I flick the VFO up the band and there he is. Even better, no pile-up! I give a quick call and he's in the log. From getting the information to the smug satisfaction of working a new one all inside 20 seconds! Just as good, after I've worked him a couple of the other local lads from the cluster call in and work him before the wolf-pack descends!

"As I tune away I notice an interesting sounding accent a few kilohertz up. Let's see now. Ah, it's Jim, VK9NS! I'm sure that someone's going to need that! So, I enter a simple command:

```
       DX 18158.4 VK9NS Norfolk Island
```

A couple of seconds or so later that information is making its way into the shacks of the other DXers logged onto the cluster.

"Ah, that's nice – as I tune away I hear a couple of the other lads calling and working him. Co-operation at its best!

"In a pile-up on a big expedition the cluster comes into its own as well. When the 3Y5X expedition was on it was often hard to tell exactly where the operator was listening but, using the cluster, we were able to keep track of where their RX VFO was, with a number of people tuning up and down the band spotting the stations that they were working. By inputting this information into the cluster it was possible for those who still needed a QSO to have a good idea where to place their own TX VFO for best effect.

"What other information can we get out of the cluster? When we first come into the shack in the evening it's nice to get a picture of what's happening on the various bands. So after signing onto the cluster I can check on what's been happening over the last few minutes:

```
       7004.12  UJ8JKK     17-Sep-1990  1958Z        <GW3YDX>
       14258.6  V63AA      17-Sep-1990  1946Z        <GW4BLE>
       14165.0  FK8FA      17-Sep-1990  1944Z        <GW0ARK>
       14165.0  FK8FA      17-Sep-1990  1944Z        <GW4BLE>
       14165.0  HZ1HZ      17-Sep-1990  1938Z        <GW5NF>
       G4VXE de GB7DXC     17-Sep 2048Z >
```

Hmm, looks like the SSB lads are having a field day with the FK8FA operation. I'm not interested in that though, so let's see what's been going on 18MHz:

```
       18081.0  EA8BWW     17-Sep-1990  1910Z        <G3FXA>
       18075.0  9Y4VU      16-Sep-1990  2121Z        <G3NOH>
       18161.0  VP8CBL     16-Sep-1990  2119Z        <G3XTT>
       18084.9  PZ1DV      16-Sep-1990  2112Z        <G3NOH>
       18154.2  V47NXX     16-Sep-1990  2110Z        <G4VXE>
       G4VXE de GB7DXC     17-Sep 2049Z >
```

Okay, well there's not been much on tonight, but it's early yet. The VP8 looks interesting though. Wonder if he'll be on tonight. I could easily check on his operating patterns and see when he is usually about.

```
       18161.0  VP8CBL     16-Sep-1990  2119Z        <G3XTT>
       18133.0  VP8CBL     12-Sep-1990  2124Z        <GW4BLE>
       18160.0  VP8CBL     11-Sep-1990  2141Z        <GW4BLE>
       18150.0  VP8CBL     11-Sep-1990  2117Z        <GW0ANA>
       G4VXE de GB7DXC     17-Sep 2127Z >
```

"That's useful! It looks like the time to listen for him is between 2115 and 2145z somewhere between 18,130 and 18,160. It might be worth checking out whether or not we'll have any propagation – I haven't checked the solar flux myself tonight. Fortunately, G3COJ and G3XKD do a good job of putting in the solar figures most days. Let's see what the figures are today:

```
Date         Hour   SFI   A    K    Forecast
17-Sep-1990  18     205   10   2    SOLAR LOW/MODERATE, GEO. UNSETTLED
16-Sep-1990  18     200   22   4    SOLAR LOW/MODERATE GEO ACTIVE
16-Sep-1990  06     203   21   3    SOLAR MODERATE, GEO ACTIVE
15-Sep-1990  06     206   19   4    SOLAR MODERATE, GEO. UNSETTLED
14-Sep-1990  06     194   20   4    SOLAR LOW/MOD, GEOM ACTIVE
G4VXE de GB7DXC     17-Sep 2049Z >
```

That looks pretty promising – the flux is on the way up and the A and the K indices aren't high enough to suggest that we'll have a fade out on our hands. The same information is really handy for keeping an eye out for auroras which can liven up 50 and 144MHz for some DX propagation.

"What's that on the other receiver? Ah, W1CDC/8R1. I worked him a couple of days back, but I think a couple of the lads still want that one. Perhaps I'll enter it into the system just in case they're around and fancy giving him a call. I feed the information in the same way as before to give them the benefit of my receiver. Anyway, it's back to the VP8 for me...

"Let's see what the cluster thinks that the MUF will be down to VP8:

```
So-Georgia propagation:    MUF: 27.4 MHz    LUF: 4.0 MHz
So-Orkney propagation:     MUF: 28.0 MHz    LUF: 4.6 MHz
So-Sandwich propagation:   MUF: 28.2 MHz    LUF: 2.3 MHz
So-Shetland propagation:   MUF: 30.0 MHz    LUF: 6.0 MHz
Falkland-Is propagation:   MUF: 27.2 MHz    LUF: 6.9 MHz
G4VXE de GB7DXC     17-Sep 2128Z >
```

No problems there, looks like if he's on 18MHz I might even have a shot at working him on 24MHz too if I could persuade him to QSY. We'll see! Since I've got a few moments I might as well find out where I need to point my beam. When I first started using the DX cluster it asked me to enter my latitude and longitude so it's now able to compute the beam heading and distance from my QTH to any of the prefixes recorded in the database on the computer. Let's see now:

```
VP8 So-Georgia:    201 degs - dist: 7583 mi, 12204 km
VP8 So-Orkney:     203 degs - dist: 8102 mi, 13038 km
VP8 So-Sandwich:   195 degs - dist: 7661 mi, 12329 km
VP8 So-Shetland:   208 degs - dist: 8445 mi, 13592 km
VP8 Falkland-Is:   215 degs - dist: 7883 mi, 12687 km
G4VXE de GB7DXC    17-Sep 2128Z >
```

That's fine... so now it's just a matter of sitting down and tuning across the band to see if he shows up. And if he doesn't tonight, well I'll know when to check the band tomorrow!

"Those are the most useful features of the cluster software for the DXer. Needless to say, there are many other lesser facilities. For example, there's a useful TALK mode. This enables me to converse with a single user or make an announcement to all the users connected to the node or to the whole DX cluster system. This can be very useful:

```
To ALL de G4VXE: Anyone know what the pile-up is on 10.104 ?
```

A mailbox is built into the software as well, so it's very easy to leave a message for a friend or to check out the latest rumours on the Albanian expedition.

"Of course, the real beauty of the system is that the more people that use it, then the more information in the form of DX spots that flows through it. And because your local node is linked to other cluster nodes around the country you can benefit from very many users with a whole range of interests: CW or SSB DXing, IOTA, 50 or 144MHz (yes, the cluster network is useful for VHF DXing too!).

"In effect then, what we all have is a great range of antennas and receivers all searching the bands, not forgetting of course that *you too* form an essential part of the system input.

"What more can I say? Using a DX cluster won't turn you into a great DXer overnight, but it'll certainly help you to be around in the right places and it's also a great place to learn the 'tricks of the trade' from some of the finest DXers in the UK.

"The DX PacketCluster is a system devised (and software written) by Dick Newell, AK1A, a professional software engineer and an avid contester and DX chaser.

"To co-ordinate and encourage clusters in the UK, the UK Cluster Working Group has been formed. It consists of members from the various clusters active at present:

    GB7DXI, Wokingham, sysop G4LJF
    GB7DXC, Cheltenham, sysop G4PDQ
    GB7YDX, Wetherby, sysop G3VMW
    GB7WDX, Crediton, sysop G3HTA

(Clusters are also planned at Burgess Hill, near Brighton, sysop G3VKW, and Hemel Hempstead, sysop G3OUF.)

"Any of the above sysops or the UK Cluster Working Group's publicity officer, George, G3LNS, would be more than pleased to provide further information regarding the clusters.

"One of the objectives of the UK Packet Working Group is to assist in the enhancement of the UK Packet Cluster Network, which should bring benefits not only to the cluster users, but also to the packet community at large."

Epilogue

It is a fascinating fact that no matter how clever a bit of computer kit becomes, it is accepted without question within a very short space of time. It might be said that it is the same with packet radio. Relying as strongly as it does on the computer fraternity, AX25 seems to be one of those applications where its smartness leads to acceptance by the user. No sooner have you connected to your first packet station and marvelled at the miracle of modern communications than you are off in search of ever more astounding feats of communicative magic. Pretty soon the mailbox will be little more than a chalk board for messages whilst a quick hop from station to station, down the country, marks a personal victory for your conquest of this latest mode. But don't be misled by the ease with which this can be accomplished. There is a lot going on 'under the hood', and much more to come. No doubt you will still long for that turbo-assisted all-mode rig that promises to span the earth with your signal, but in a startling reversal you can now communicate through the packet radio network with the most modest of equipment.

Consider if you will for a moment, a network. It can be defined in its broadest terms as a collection of nodes, each joined by some means to its neighbour. Until very recently, this meant for radio amateurs that a number of stations would all operate on the same frequency. The unenviable task of 'net controller' was almost paramount in preventing the chaos which could (and often did), arise unbidden if the 'protocol' of network operating was forgotten by any one amateur.

Using the radio as the gateway and the ether to connect the nodes of a network is not without its problems. The uncontrolled babble of a multitude of network users is useless, but the polite etiquette of conversation on a computer network will govern the flow of information as competently as the most experienced 'networker' on an evening net. At a time when frequencies are in short supply (and those we have being sniped at by commercial pressures), here is a mode which can use just one, even at the expense of time. When interference prevents many a conversation, here is a mode which thrives on perseverance.

For the moment, packet radio is growing up quickly. The final layers of the AX25 protocol are not yet in place, although enough of the groundwork has been done to allow those of us with neither the understanding nor the desire to get involved, to benefit by using and exploring its capabilities. As long as your rig can put out a signal which will reach your neighbour, you can join and become part of the network which may encompass anything from a few local stations to links via satellite. Packet may not be the sort of mode for spending an evening in the shack if only because it can be a mite slow. However, as an alternative to listening to the local chitchat it is excellent. You really can get on with building that project while the station keeps an eye on the traffic. Monitoring data (for that is surely what we do with a casual ear on the local frequency) from either voice or packet is a thankless task, but one which can now be performed by the ubiquitous computer.

Many would rather have a computer tell them the date of the next rally and who will be going than have to listen to countless similar conversations to glean the information. In the very near future the computer can monitor these disparate sources for us, whilst we get on with more pressing tasks. We are very close now!

Getting information from data is but one aspect of communications, but conveying accurate and timely data communicates information. For many, packet will never hold any fascination. The joy of speaking to other like-minded individuals, whether via a microphone or in the universal language of the morse key, may be said to be the very essence of amateur radio. But in the crowded frequencies of today's amateur allocations it is refreshing to find a mode of communication that neither requires, nor benefits from, enormous aerials or expensive rigs. It is a beautiful paradox of this science that the antisocial habits of those who would spoil our hobby can be totally ignored by the very mode of packet radio.

So that's about it to begin with. Although it seems quite bewildering at first sight, operation could not be simpler once you have got your feet wet. If the facilities are there, why not have a go? If you can drive a TNC, that's all there is to it.

*** DISCONNECTED.

G8UYZ
G8NZU

Appendix 1

TNC-2 and clone command list

Although you should find a description of all your TNC commands in the manual, we thought it might be handy to have a reference list. This list (which is mostly made up of the standard TNC-2 command set) has been collected from various sources, not the least being via the packet network itself. Your own particular TNC may not feature all of these (although it probably will), and it may have several commands not in this list. However, it represents the basic commands you should have available. Thanks are due to: KA2UGQ, N1CQE, DG3SAJ, G6UDM, and others too numerous to mention.

Note: in the case of commands which switch between two possible states, the top one shown here is the default. Where there is a numerical parameter, the one shown is the default.

Command	Value	Description
8bitconv	**OFf**	High-order bit stripped in CONVers mode.
	ON	Top bit not stripped.
AUtolf	**ON**	<LF> sent to terminal after each <CR>.
	OFf	<LF> not sent.
AWlen	**7**	Seven data bits per word to terminal.
	8	Eight data bits per word to terminal.
AX25L2V2	**ON**	AX.25 LEVEL 2 VERSION 2.0.
	OFF	VERSION 1.0.
ASyrxovr	**0**	Data from user to TNC dropped if positive (0-65535) (always 0).
AXDelay	**0**	Wait time in 10ms intervals in addition to TXdelay (0-180).
AXHang	**0**	Audio repeater 'hang time' in 100ms intervals (0-20).
BBfailed	**0**	Number of times bbRAM (battery backed RAM) check sum was in error.

Mailboxmsgs	OFf	Changes message display.
	ON	Newline added to message before "***".
BAud		Check and set baud rates.
Beacon	Every 0	Send beacon every (n*10s) intervals (n = 0-250).
	After n	Send beacon once after (n*10s) of no activity (n = 0-250).
BKondel	ON	Backspace echoed in response to <Delete> keystroke.
	OFf	<\> is echoed.
BText		Defines text to be sent as Beacon (120 characters maximum).
BUdlist	OFf	Ignore packets from stations IN LCAlls.
	ON	Ignore packets form stations NOT in LCAlls.
CALSet	n	Count parameter for calibrate.
	z	n = (525,000/f)+1 – mod.
		n = (262,500/f)+1 – demodulate.
CANline	<00-7F>	Character used to cancel input, usually $18, ie ^X.
CANPac	<00-7F>	Character to cancel packet, usually $19, ie ^Y.
CBell	OFf	Disable connect bell on terminal.
	ON	Enable connect bell.
CHeck	30	Time in 10s intervals to check for disconnect (0-250).
CLKADJ	0	Correction factor for real-time clock (0-65535).
CMdtime	1	Transparent timeout in seconds – used for 'break' (0-250).
CMSg	OFf	CText not sent.
	ON	CText sent after connection by another TNC.
CMSGDisc	OFf or	
CTEXT	empty	Forces immediate disconnect after CText.
	ON	Send CTEXT before DISC.
COMmand	<00-7F>	Character to enter command mode from CONVers, usually $03, ie ^C.

CONMde	**CONV**	Converse Mode entered automatically on connect.
	TRAN	Transparent mode.
Connect	**<call>**	Establishes link.
		Also C <call> via <call2, call3, ... >.
CONPerm	**OFf**	Current stream is temporary and may be connected to and disconnected from.
	ON	Current stream is permanent and may not be disconnected.
CONOk	**ON**	Connect requests accepted from other TNCs.
	OFf	Requests not accepted.
CONStamp	**OFf**	Connect status messages not time stamped.
	ON	Messages are time stamped.
CONVers		Enter CONVers mode from Command mode (alternate command – K).
CPactime	**OFf**	Packets not sent according to PACTime in CONVers mode.
	ON	Timed packet dispatch is used.
CR	**ON**	SEndpac character sent along with packet.
	OFf	SEndpac character not sent.
CStatus		Displays stream id and link status of all streams. (...P indicates permanent).
CText		Text sent on Connect if CMSg ON (120 characters maximum).
DAytime	**<yymmddhhmm>**	Set current date and time in 24 hour format.
DAYUsa	**ON**	Date display USA format: mm/dd/yy.
	OFf	Date in European format: dd-mm-yy.
DELete	**OFf**	Delete character is <BS> or $08.
	ON	Delete character is <DEL> or $7F.
DIGipeat	**ON**	TNC will digipeat packets on request.
	OFf	Digipeat disabled.
DIGISent	**0**	Frames digipeated by TNC since power-up or restart (0-65535).

Disconn		Initiate Disconnect sequence (A second 'D' makes disconnect immediate).
DISPlay		Display all control parameters and their current values:

          A sync   Serial port parameters.
          C haracter   Special characters.
          H ealth   Counter values and HEALled status.
          I d   ID parameters.
          L ink   Link parameters.
          M onitor   Monitor parameters.
          T iming   Timing parameters.
          A LL   All values (default).

DWait	16	Wait time in 10ms intervals after last activity (0-250).
Echo	ON	Characters received by TNC are echoed.
	OFf	Characters are not echoed.
EScape	OFf	<ESC> character output as $1B.
	ON	<ESC> character output as $, ie $24.
Flow	ON	Terminal input stops TNC output temporarily.
	OFf	TNC output unaffected by terminal.
FRack	3	Frame acknowledge wait (1-15s).
		Retry (seconds) = n*(2*[# relays]+1).
FUlldup	OFf	Full duplex mode disabled – DCD used.
	ON	Full duplex enabled.
HEaderln	OFf	Header on same line as monitored packet text.
	ON	Header on separate line.
HEALled	OFf	CON & STA LEDs show normal functions.
	ON	CON & STA LEDs flash to monitor TNC software.
HOvrerr	0	Data lost if HDLC RX not serviced (0-65535) (always 0).
HUndrerr	0	Frames aborted if HDLC TX not serviced (0-65535) (always 0).
HId	OFf	No ident packet (in digi-on mode).
	ON	ID packet sent every 9.5 minutes if station is digipeating.
Id		Sends special ID packet from command mode.

TNC and clone command list

K		Equivalent of CONVers (undocumented).
LCAlls	<call1...call8>	List of calls for BUdlist.
LCok	ON	Lower case characters sent to terminal.
	OFf	Only upper case characters sent.
LCStream	ON	Character following STReamsw char is not significant.
	OFf	It is significant (immediate process).
LFadd	OFf	<CR> character only sent in packet.
	ON	<LF> added to <CR>.
LFIgnore	OFf	TNC will respond to <LF>.
	ON	<LF> ignored).
MAll	ON	All monitored packets displayed.
	OFf	Only Unproto and status packets displayed.
MAXframe	4	Number of unacknowledged packets allowed (1-7).
MCOM	OFf	Only data packets monitored.
	ON	C/D/UA/DM frames also monitored.
MCon	ON	Monitor mode enabled when TNC is connected.
	OFf	Monitor mode disabled.
MFilter	<00-7F>	Up to 4 characters to be filtered by TNC.
MHClear		Clears MHeard list.
MHeard		Lists last 20 calls heard, with date ('*' indicates heard through digi).
Monitor	ON	Enables general packet monitoring.
	OFf	Disables monitoring.
MRpt	ON	Display path for monitored packets.
	OFf	Show origin/destination only.
MStamp	OFf	Monitored frames are not time stamped.
	ON	Each frame is time stamped.
MYcall	<call>	Your callsign (optional -n where n is 0-15).
MYAlias	<call>	Alternate callsign for digipeater (optional -n where n is 0-15).

NEwmode	**OFf**	Switch to CONVers/Trans on connect only.
	ON	Switch occurs at time of Connect command and returns to cmd: after disconnect.
NOmode	**OFf**	Mode switching determined by NEwmode.
	ON	Mode switching by user only.
NUcr	**OFf**	Null chars not sent to terminal after <CR>.
	ON	Null characters sent.
NULf	**OFf**	Null chars not sent to terminal after <LF>.
	ON	Null characters sent.
NULLs	**0**	Number of nulls to sent after <CR> or <LF> (0-30).
PAClen	**128**	Maximum length of data part of packet (0-255, 0 = 256).
PACTime	**After n**	Input bytes sent after n*100 ms with no input.
	Every n	Input bytes sent every n*100 ms. No zero length packets (n = 0-250)..
PARity	**3**	Parity bit setting to terminal (0 or 2 = none; 1 = odd; 3 = even).
PASs	**<00-7F>**	Character after this sent 'as is' and not interpreted as a command, usually $16, ie ^V.
PASSAll	**OFf**	TNC only accepts packet with valid CRC.
	ON	Accepts any packet.
REConnect		Change connection path while connected (arguments as for Connect command).
REDispla	**<00-7F>**	Character to redisplay last input line; usually $12, ie ^R.
RCVDIfra	**0**	I frames received (0-65535).
RCVDRej	**0**	REJ frames received (0-65535).
RCVDSabm	**0**	SABM frames received (0-65535).
RESET		Resets all parameters to default settings.
RESptime	**5**	Minimum delay (n*100ms) until acknowledge packet sent (0-250).

RESTART		Reinitialises TNC to user default settings in battery-backed RAM.
REtry	10	Maximum number of packet retries (0-15: where 0 = infinite retries).
RXblock	OFf ON	Data sent to terminal in standard format. RXblock format.
RXErrors	0	Frames rcvd with bad CRC, too short or too long (0-65535).
ScreenIn	80	Sets screen width of terminal (0-255: where 0 = disable).
SEndpac	<00-7F>	Character to force packet to be sent, usually $0D, ie ^M.
SENTFrmr	0	Frame reject frames sent (0-65535, usually 0).
SENTIfra	0	I frames sent (0-65535).
SENTRej	0	REJ frames sent (0-65535).
STArt	<00-7F>	Character to restart output from TNC after STOp, usually $11, ie ^Q.
STOp	<00-7F>	Character to stop output from TNC to terminal, usually $13, ie ^S.
STREAMCall	OFf ON	Callsign of other station not displayed. Callsign displayed.
STREAMDb	OFf ON	Do not double RECEIVED streamswitch chars. Double streamswitch chars.
STReamsw	<00-7F>	Stream switch character, usually $7C, ie |.
TRACe	OFf ON	Disable hex trace mode. Display trace mode with hex, shifted ASCII and ASCII.
Trans		Enter transparent mode.
TRFlow	OFf ON	Use hardware handshake for terminal. Use XON/XOff in Transparent mode.
TRIes	n	Current number of REtry level (0-15).

TXCount	0	Frames transmitted correctly (0-65535).
TXdelay	30	Wait time after key up before sending packet; n*10ms (where n = 0-120).
TXFlow	OFf ON	Use hardware handshake for TNC. Use XON/XOff in Transparent mode.
TXQovflw	0	Number of frames discarded because outgoing frame queue too small.
TXTmo	0	Count of successful recovers from HDLC transmitter timeouts.
Unproto	call1 TO	<via call2[,...call9>. Address for unconnected packets (CQ).
USers	1	Set number of active input connections allowed (0-10: 0 = number 10).
Xflow	ON OFf	XON/XOff flow control enabled. disable XON/OFF flow control.
XMitok	ON OFf	Transmit enabled. Transmit disabled.
XOff	<00-7F>	Character output by TNC to stop terminal input, usually $13, ie ^S.
XON	<00-7F>	Character output by TNC to start terminal input, usually $11, ie ^Q.

Appendix 2

Packet mailbox commands

Even the best-intentioned packet users never quite get around to downloading the mailbox manual, and there always seems such a lot to print out... With due acknowledgment to Andy Witts, G1DIL, (who was instrumental in getting many a packeteer on-air in the West Midlands) we hereby save you some trouble.

This guide is intended primarily for those not *au fait* with the mailbox, and who need a bit of documented assistance. Seasoned users may find it useful so that they are operating as efficiently as possible. There will from time to time be amendments or new versions which may change the list; but at least it will help you out in the early stages. In short, if you are a packeteer, please allow five minutes to read the commands below, and keep a copy by the side of your TNC.

It is often the intention of the authors of mailbox software to continually update the services and facilities available to users. It is naturally anticipated that new versions will be available from time to time, and as such some of the commands and their functions may change. This, unfortunately, is the price one pays for progress; however, the basic commands will probably not be changed. They may, however, be fine-tuned, and as such the writers assume no responsibility for errors that may occur from current or future versions of this or any other mailbox software.

Now you are ready to explore the mailbox. On the following pages are documented most of the commands for a variety of mailboxes. We have included the WA7MBL box (written by Jeff Jacobsen), as most mailbox software commands are based upon this set and the explanations are more or less common.

The WA7MBL BBS

A – abort
If you find the mailbox sending you information you do not want, type **A** at any time. This will abort the procedure and return you to the menu with the statement:

```
*** Aborted by user ***
```

Note that this will stop data flow from the host to the TNC only; you are committed to take what data is already in the TNC buffer.

B – bye

The 'B' command logs you off the mailbox and disconnects your call. Type B to log off gracefully and reset the system for the next user. Note: you may also log off by disconnecting, however, a system reset is not 100% guaranteed within one minute.

Bulletins

A bulletin is a message flooded to either *every* mailbox in the network, or to *selected* mailboxes by using a route qualifier. Each bulletin is issued with a bulletin identifier (BID), which ensures that the message is unique and so can be passed efficiently around the network. To send a bulletin, first get used to sending personal mail (read the section 'S – Sending Mail'), and then add the qualifiers. Note that the assignment of BIDs is automatic and so should not be entered by the user. To send a bulletin to *every* mailbox use the syntax:

    S [qualifier] ALL $

The '$' requests the mailbox to assign a BID (bulletin ident). To send a bulletin over a selected route, use the syntax:

    S [qualifier] [route info] ALL $

For example, to send a general information bulletin use:

    SB ALL $

whereas sysops can send messages to all other MBL operators by sending:

    SP SYSOPS @MBLMailbox $

Note: bulletins create immense amounts of auto-forward traffic and so should be used with care, so as not to overload the network. Consult your local sysop for selective routing information in your area.

D – download

The 'D' command lets you download files from the mailbox text file section. You have a number of options when downloading files, to ease traffic flow. First establish if the file you wish to download is in the main or sub-directories. If it is in the main directory then use the syntax:

    D name

If the file is in a subdirectory then you must specify the download path by using the syntax:

```
D path\name
```

So to download a file entitled 'Radio' which is in the GENERAL directory:

```
D General\Radio
```

A most useful feature of the mailbox text files section is the VIEW facility, which allows you to download only small sections of the file. To use this type:

```
D name 10
```

to view the first 10 lines of the file;

```
D name 10 20
```

to view lines 10 to 20;

```
D name -20
```

to view all lines after 20. Note: you cannot 'download' mail. Use the 'R' command to READ messages.

DU – display user

You may interrogate the user database at any time to see the details of any particular station. This is done using the syntax:

```
DU callsign
```

You will then be shown:

```
Date/Time      Call    Name LstMsg #    TNC
8801011200     G1DIL   Andy 10540       157 A
```

Files

If you have an interesting file of your own or a non-copyright (public domain) program, please upload it to your mailbox. You may see subdirectories which may have been set up to house items of similar interest. For example there may be a subdirectory set up for BBC-B users or similar as well as one for the PC. Check before uploading if your item would be best suited to the main or a subdirectory. Use of these subdirectories will speed location of specific files by other users.

Note: you must use YAPP protocol for up/downloading binary files, which is explained elsewhere in this guide. Please check available disk space prior to uploading to ensure there is sufficient storage capacity available.

H – help

When stuck for the next command, users may key H and will then be presented with a typical short mailbox command set. But beware; if you ask to see this too often during a session, you will be disconnected.

```
B...Log-off
D...Download
DU call..status
I...Commands info
J...MH Lists
K...Kill Mail
L...List Mail
N...Name input
R...Read Mail
S...Send Mail
T...Talk to SysOp
U...Upload
V...Lst Msg #
W...Directory
X...eXpert toggle
```

I – information

Sends a further information file to expand on the above menu – if required.

K – kill mail

The 'K' command allows you to kill (delete) your mail. You can only kill mail that is either to or from you; you cannot kill other people's mail.

Note: you can usually only kill mail that you have already read.

Use the syntax:

```
KM
```

to kill all mail you have read, ie Kill Mine, or

```
K 1234
```

to kill a specific message number (up to six per entry). You should always kill your mail prior to logging-off. Make this a habit.

Note: only the sysop can kill text or binary files, so if there's one of yours you would rather not see on the board any more, ask the sysop to kill it for you.

L – list mail

The 'L' command is the most versatile command. It can be a superb aid to operation if used correctly. Used incorrectly it will cause you and your local packeteers much congestion. The 'L' and its qualifiers list mail which is currently on-file. If you give the 'L' command on its own it will list *everything* since the *last time* you gave this command. It makes sense to give this command regularly. The command can be qualified using the following:

L	to list everything since last issued.*
LA	to list Amsat bulletins.
LB	to list general bulletins.
LI	to list TCP/IP Internet bulletins.
LM	to list all mail addressed to you.
LN	to list new mail addressed to you.
LQ	to list GB2RS/RSGB Bulletins.
LR	to list Raynet mail.
LW	to list 'wanted' bulletins.
LX	to list Packet Working Group bulletins.
LL 10	to list last 10 messages (variable).
L 150	to list all after msg 150 (ditto).
L 150 300	to list all between msg numbers.
L>Call	to list all to that callsign
L<Call	to list all from that callsign
L@ Call	to list all with @xxx routing information

*Use with caution: this list could be long!

Most mailboxes will give you an indication of the number of messages that you will see when giving the 'L' command, when you log on. It is advisable to read it.

N – name input

Type **N** to enter or change your name.

R – read mail

The 'R' command lets you read messages. Use the following syntax:

RM	to read *all* mail addressed to you.
RN	to read *new* mail addressed to you.
R 1234	to read message number 1234.
R 1234 4321	to read two, or more specific messages, up to a maximum of six.

Note: when reading mail using the 'R' command, any forwarding headers are removed to reduce transmission time. Should you wish to view the list of headers, read the message using the 'V' commands, which are shown later. R ALL and READ ALL are not valid commands – use the list command.

REQDIR

If you wish to receive a directory listing from a distant mailbox, please use this command rather than attempting a DX access. Address your message as follows:

    S REQDIR @ GB7xxx

Then in the title you should give (a) the directory information, and (b) your 'home' address. (Note that wildcarding (*) is permitted.)

So if you are at GB7HQM and wish to see the ROOT or MAIN directory at GB7xxx you would enter the following in the title field:

    . @ GB7HQM

where '*.*' is the wildcard for all files. If you wished to see a listing of the BBC sub-directory you would send:

    \BBC @ GB7HQM

having shown the path information to the sub-directory. These messages are returned from the distant mailbox automatically during the next auto-forwarding session.

REQFIL

Having established from the above procedure which file you would like to receive, and in which directory it resides, you should now send the command to request auto-forwarding of the file. Address the message as follows:

    S REQFIL @ GB7xxx

Then in the title you should give (a) file and path information, and (b) your 'home' address. So if you require a file called MAILBOX.USE which is in the ROOT or MAIN directory at the distant mailbox you would enter:

    MAILBOX.USE @ GB7HQM

If you wished to receive a file called TINY2.MOD which is in the GENERAL subdirectory, you have to specify the filename *and* the path information as follows:

    GENERAL\TINY2.MOD @ GB7HQM

The file will then be sent to your 'home' mailbox addressed to yourself, to read as a message during the next forwarding session.

S – send mail

The 'S' command allows you to send mail to other users. Use the syntax below:

S [Callsign]	to send an open message.
SB [ALL]	sends a bulletin to everyone.
SP [Callsign]	to send private mail.
SW [ALL]	ends a 'wanted' bulletin.

See other mail categories shown under the 'List' command for type qualifiers. *Always use category qualifiers when sending bulletins.*

Note: you can substitute [Callsign] for [ALL] in any of the above examples.

Once you have entered the 'S' command string with the callsign, you will be prompted for:

(i) The title. Be sure to use a title which is meaningful and not just one word like 'Help' – help with what? Please give each title some thought.
(ii) The text. Be sure to end your text with ^Z or /EX (on a new line) to terminate input to the disk.

To send mail to users at other mailboxes use the 'S' commands as shown above but now add the callsign of their nearest mailbox. So to send mail to G3OUF whose 'home' mailbox system is GB7HQM, send:

    SP G3OUF @ GB7HQM

When sending international mail use the syntax shown above, for example:

    SP G5ABC @ GB5XYZ

and let the mailbox do the routing for you. You *must*, however, know the callsign of the recipient's 'home' mailbox to ensure delivery. (Most mailboxes carry information regarding mailbox lists for popular international destinations.) If you do not know the destination mailbox callsign *do not send it*, as it will only serve to choke the network, at least until the latter is more advanced.

SERVER y

Address a message to Server @ GB7xxx and, provided you request information in the correct format, the mailbox will generate a reply (in non-real-time) with the calculated information. This is useful for such tasks as orbital calculations, QRA calculations and the like.

You should consult your local sysop to see: (a) what SERVER facilities are available on your 'home' mailbox; (b) what SERVERs are available at distant mailboxes; (c) what format your enquiry message should take.

T – talk to sysop

Type T to page the sysop. If this facility is available, you will receive a reply within one minute, if not you will be advised of further action to take.

U – uploading

The 'U' command lets you upload a file to the mailbox text files section. Type:

    U filename.extension

to upload a file into the **M**ain directory. For example, type U `RADIO.TXT` to upload the file RADIO.TXT. To upload files into a sub-directory you have to specify the path:

    U `path\name`

For example, to upload the file named 'Oscar12.kep' into the AMSAT sub-directory, you would type:

    U `\Amsat\Oscar12.kep`

Note: you cannot 'upload' mail. Use the 'S' command to send mail.

V – verbose
The 'V' command lets you read messages together with their forwarding headers (unlike the 'R' command which strips all headers) in the 'verbose' form.

VM	to read all mail addressed to you.
VN	to read new mail addressed to you.
V 1234	to read message # 1234.
V 1234 4321	to read two, or more, specific messages, up to a maximum of six.

V – version
Sending the command 'V' alone to the mailbox will allow you to see the status of the system. This information will include the type and release number of the software in use, the last message number issued by the machine and the number of active messages at that time.

W – what's new?
The 'W' command lists files available on the disk. Files are divided into sub-directories by class or category to speed location.

W	to list the MAIN directory
WN	to list new files in the MAIN directory.

'New' means since the last time you catalogued the disk. There will also be sub-directories, which will appear as, for example, 'AMSAT' when you do the directory listing. To list files in these directories, type the sub-directory name after the 'W'.

W AMSAT	to list files in the Amsat sub-directory.
W BBC-B	to list the directory used by BBC-B owners.

To download the files, use the commands shown above under the 'D – downloading' command.

X – expert status
Type **X** to toggle between expert and non-expert. If you have been seen to use the mailbox regularly, the sysop may change your status for you.

Yfiles (binary files)
The 'Y' commands call the YAPP protocols with which you can upload and download binary files. You *must* use the YAPP packet terminal program in *your* computer – binary transfers with 'MBL mailboxes will not work with any other terminal programs or emulators. At present, the YAPP program is available *only* for IBM-PC compatible PC/MS-DOS computers and the BBC Micro.

The **'YW' command** shows 'What's New' in the binary files area of the host disk. Type YW to see a listing of the binary ROOT directory. The mailbox will send its directory of available binary files in 'wide-screen' MS-DOS format, showing the individual file sizes and dates, as well as the disk space available at the host.

The **'YN' command** shows any new files added to the ROOT directory. Type YN to see what new binary files have been added to the ROOT directory since you last logged on. File dates and sizes are also shown.

The **'YU' command** allows users to upload binary files to the mailbox. Type YU filename.extension to upload a file. (If you omit the filename, the mailbox will prompt you with Enter file name.) The mailbox will respond with: Ready to Receive with YAPP protocol. Be sure to include all necessary path and filename information so that YAPP can locate your file for uploading. For example, to upload the binary file 'PACKET.COM', proceed as follows. At the normal command prompt, type:

    YU PACKET.COM

The mailbox responds with:

    Ready to Receive with YAPP protocol

Press the <F7> (Bin Send) key. A small window will open on your screen and prompts you to:

    Enter Filename:

type PACKET.COM and your keystrokes will appear in the small window.

In a few seconds, the upper-right area of your screen will show activity. Your TNC will automatically be switched into Transparent Mode and the binary transfer will begin. The small window will be replaced by a larger window which now shows

YAPP File Transfer Protocol status and progress data. The new window shows you the name of the file being uploaded, the size of the file in bytes, the state of the transfer, the number of bytes transferred and the `Press to abort` prompt.

During the file transfer, the state information will change to show you what is happening between your computer and the host.

When the entire file has been transferred, the window will show:

```
SEND COMPLETE - Awaiting Final Ack
```

and then:

```
Transfer completed OK
```

and

```
Returning to Terminal Mode
```

The window will close and the mailbox will send you its normal command prompt indicating that it is ready for your next command. If your 'upload' is large, be sure to use the 'YW' directory command to check available disk space before sending the file.

The **'YD' command** allows users to download binary files from the mailbox. Take the time to verify the *correct* filename of the file you wish to download, along with the path information.

Type YW to see the directory of binary files. The mailbox will send its directory of binary files, show you the size of the file you want, and the disk space available. Type `YD filename.extension` to download a file. The mailbox will respond with `Ready to Send with YAPP protocol`. For example, to download the binary file 'PACKET.COM', proceed as follows. At the normal command prompt, type:

```
YD PACKET.COM
```

The host responds with:

```
Ready to Send with YAPP protocol
```

Press the <F8> (Bin Recv) key. A small window will open on your screen and prompts you to:

```
Enter Filename:
```

Type `PACKET.COM` (or whatever path and filename); your keystrokes will appear in the small window. The rest of the procedure is as described above for uploading binary files.

General information
A user may obtain brief instructions for any system command by typing ? followed by the command letter. For example typing ?L will cause the mailbox to send the on-line help file relating to listing mail headers.

The W0RLI mailbox
Note: much of these notes were originally culled by G1DIL from those supplied with the ROM itself, and subsequently edited by the authors. Several were collected 'on the fly' as it were, by typing the 'H' followed by whatever, so these instructions are as obtained from the screen of a computer one night. After this information, we have included the help text taken 'off air'.

For help on a specific command, enter H x where 'x' is the command for which you wish help. For example, H R will give help for the READ command. Optional fields of the commands are shown inside '[]'. Note that this may not be the way it is done on all boards at the moment.

Message commands
(K)ill
(L)ist
(R)ead
(S)end

File commands
(D)ownload
(U)pload
(W)hat

GateWay commands
(C)onnect
(M)onitor
(RT) RoundTable

Misc commands
(H)elp (?) Same as H
(B)ye
(I)nfo
(J) Who?
(N)ame
(T)alk to sysop
(V)ersion

H	Gives a summary of the commands.
H x	Gives an explanation of command x. Example: H R will give details of the Read command.
H *	Gives an explanation of all commands.
B	Log off the mailbox. (Disconnecting has the same effect.)

Cp call	Connect to call using port p. Example: `CA G9XXX`.
CM # call	Copy message # to call. Same syntax as S command. Example: `CM 123 G9XXX`.
Dd filename	Download a file from the mailbox where 'd' is the path identifier. Example: `DA DESIG.NTS`
I	Gives a paragraph on the hardware, software and RF facilities of this mailbox station.
IL	List local users of this mailbox.
IZ ZIP	List users at ZIP code. Trailing '*' as wildcard.
Jp	Here p is a port identifier. Gives a short list of stations recently heard on that port. The console port list shows the calls of stations recently connected to the mailbox.
K #	Kills message number #. Example: `K 327` kills message # 327.
KM	Kills all messages addressed to you, that you have read.
KT #	Kills an NTS traffic message.
L	List all new messages since your last log-in. Lists messages in reverse order, newest to oldest. 'Personal' messages not to or from you will not be listed.
L #	List all messages back to message #. Example: `L 325` lists all messages to # 325.
LL #	List the last # messages. Example: `LL 10` lists last 10 messages.
L call	List all messages to this callsign. Example: `L G9XXX` lists all messages to G9XXX.
L call	List all messages from this callsign. Example: `L G9XXX` lists all messages from G9XXX.
L@ call	List all messages addressed at this mailbox callsign. Example: `L@ GB7SAM` lists all messages at GB7SAM.
LB	List all bulletins.
LF	List all messages that have been forwarded.
LH	List all held messages.
LK	List all killed messages.
LM	'List Mine'. Lists all messages to you.
LO	List all 'old' messages.
LP	List all personal messages.
LT	List all NTS traffic.
LY	List all messages that have been read.
Mp	Monitor port p.

Packet mailbox commands 101

N xxxx	Enter your first name into user data base.
NE	Toggle your 'expert user' status.
NH xxxx	Enter your 'Home BBS' (Assists in routing messages to you.) Example: Please use the NH command to enter your home mailbox.

```
NH GB7SAM 2148z. OK Dave, Stoke Area Mbx :
(B,C,D,H,?,I,J,K,L,M,N,P,R,S,T,U,V,W)
```

NQ	Enter your QTH.
NZ xxxx	Enter your postal code. (Assists in routing messages to you.)
P CALL	Query the local WP database for all info on call. Example: `P G9XXX`
R #	Read message number #. Example: `R 325` reads message 325.
RH #	Read message number #, showing all routing headers. Example: See R above.
RM	'Read Mine'. Read all your unread messages.
RT	Invoke RoundTable, use `H RT` for more information.
S	Show system status.
S? xxxx [@ yyy]	Send message type '?' to station 'xxxx', at optional BBS 'yyy'. The mailbox will prompt for the message title and then for the message text. End text entry with a <Ctrl-Z> or /EX. '?' is an optional 'type' of message. They include: B – Bulletin, P – Personal. Only the addressee can read or list this type. The form **SB xxxx [@ yyy] [$BID]** is also available for compatibility with the WA7MBL bulletin handling systems. Example:

```
SP G9XXX @ GB7CHS SB ALL @ GB
```

T	Talk to the sysop. Any command or <Return> before the request times out will return you to the normal mailbox prompt.
U filename	Upload a file to the name given. Example: `UC WESTNET.BBS`. Reject will occur if filename already exists.
V	Show what version of the mailbox is running.
W	Gives a list of directory areas available on the mailbox.

Wd   Gives a list of the files in directory area 'd'.
Wd ffff.xxx   Gives a list of files in directory area 'd' that match the given file specification. Global commands allowed.
Using W and directory ID:

WM	aMtor users'
WA	Archimedes users'
WT	aTari users'
WX	aX-25 packet information
WB	Bbc users'
WG	General files
WI	Ibm and clones
WK	Kantronics tnc
WP	Packet working group
WR	Raynet information

Example:

```
wi HELP.BAT
10k 890505 1 files using 10k of 20810k, 11306k free.
```

Note that lower case is permitted.

My local mailbox currently runs the W0RLI software, version 11.13. The text below is taken from a recent attempt to get some help. Full details of help are not listed as it can take ages. You should be able to get by with this. All you have to do is try it out yourself.

```
c gb7max
*** CONNECTED to GB7MAX
[RLI-11.13-CH$]
GB7MAX P/C  217 0   Thank you Dave >
h ?                                    [You get this file with ? ?]

1. To receive complete help information on a specific command,
enter H  followed by a space and then the command letter.
Example: H R will tell you all about the READ command.  When you
actually use the commands, you'll find that many of them require
that something additional be entered after the initial letter,
such as a callsign or message number.  The help file will tell
you about those requirements.
2. A brief explanation of ALL the commands can be obtained by
entering ? CMD (that's ? space CMD).
3. To receive a complete description of ALL of the commands,
enter: H *  (That's H space asterisk).  Regular MailBox users
will want to do this to become familiar with all of the features
of the system.  This information fills six pages, so turn on your
```

Packet mailbox commands 103

printer or have your buffer ready to store the information before entering H *.
4. A short "Help File of Basic Commands" is available to enable new users to use the MailBox right away. It will tell you how to get a LIST of messages, how to READ messages that interest you, and how to SEND messages to others. To see this, enter ? ?
(That's two question marks with a space between them.)

To LIST messages that have been received by the MailBox since you last checked in, enter: L
To LIST recent messages, enter: LL xx (xx = the number of messages you want to list.) Example: LL 15 will list the last 15 messages.
To READ a message, enter: R and the message number.
    (Enter a space between the R and the number.)
    Example: to read message 4350, enter: R 4350
To SEND a message to another ham who uses this MailBox, enter:
    SP and the station callsign. (Enter a space between the SP and the callsign.) Example: SP G1DKI
To SEND a message to a station who uses another MailBox, enter: SP, the callsign of the receiving station, @, the callsign of the ham's MailBox, a period, and the 2 letter state abbreviation.
    Example: SP G1DKI @ GB7MAX.GBR   (Note the spacing!)
To KILL (erase) a message, enter K and the message number.
    Example: to erase message 6112, enter: K 6112
To log off the MailBox (to say goodbye), enter: B

GB7MAX P/C 217 0   Thank you Dave >
? cmd   *[Get the whole list]*

B - BYE —— Disconnect from the MailBox.
CM- COPY MSG - Make a copy of a message for another station.
D - DOWNLOAD - Download files. (Read files that are in the MailBox.)
ET- EDIT TFC - Edit the message header (TO, FROM, etc.) of an NTS message.
H - HELP —— Help in using the commands available on this MailBox.
I - INFO —— Information on the computer, software and hardware.
I -(with call) Information from the user database for that callsign.
J - WHO? —— Listing of stations recently heard or connected to the MailBox
K - KILL —— Kill (erase) messages.
L - LIST —— List messages. (Several variations available.)
N - NAME —— Enter your Name, QTH, Zip, Home MailBox into WP database.
R - READ —— Read messages.
S - SEND —— Send messages, and STATUS - Show System Status.

104   Packet Radio Primer

```
T - TALK — Talk to the sysop.
U - UPLOAD - Upload files. (Put files into the MailBox.)
V - VERSION - Find out what version of the W0RLI MailBox program
is on line.
W - WHAT — List file directories and file titles.
```

Some of the commands require added information after the command letter. For detailed information on a specific command, enter H x, where x is the command letter. Example: H L will give you information for LIST. Use the command H SERV for information on extended MailBox services.

```
GB7MAX P/C  217 0  Thank you Dave >
i                               [Get some information]
```

GB7MAX is a general purpose packet radio server node. Operating as a satellite in the UK NTS system it supports hierarchical message store and forward facilities, as well as user file upload and download. The current mailbox code is by W0RLI version 11.13, running a radio on each of 144.650 and 432.675 via G8BPQ's multi-user front end software v3.53 giving access to up to 4 callers

Located in the village of Perton some 9Km to the west of Wolverhampton town centre, the system is operated by MAXPAK, the Midlands AX25 Packet Radio Users Group, under the guidance of SysOp, Mick G1DKI.
Further information on MAXPAK can be gathered from the files area.
Or from G1NZZ @ GB7MAX   The Membership Secretary

```
GB7MAX P/C  217 0  Thank you Dave >
b
*** DISCONNECTED
cmd:
```

The AA4RE mailbox

The following was taken straight off the mailbox, and is shown complete. Some of the commands are rather like early W0RLI!

H Gives a summary of the Help Subsystem.
H x Gives a detailed explanation of command x.

For help on a specific command, enter **H x** where **x** is the command for which you need help. For example, **H R** will send complete help for the READ command.

Message commands:	(K)ill (L)ist (R)ead (S)end
File commands:	(D)ownload (U)pload (W)hat
Misc commands:	(B)ye (H)elp (I)nfo
	(J) Who? (N) Register
	(T)alk to SYSOP
Further info:	(@) At BBS

B   Log off the mailbox. Simply disconnecting has the exact same effect.

D topic ffff.xxx or DB type topic ffff.xxx

Download the file ffff.xxx for the topic requested. D is for ASCII format. DB is for binary downloads. To get a list of topics, use the W command. The types for binary download are: XMODEM, XMODEMCRC, YMODEM, YMODEM-BATCH (alias YMBATCH), and YAPP.

DU callsign   Will display user info on a specific call.

I   Gives a paragraph on the hardware, software and RF facilities of this mailbox station.

J   Gives list of ports.
Jp   Where p is a port identifier. Gives a short list of stations recently heard on that port.

JL   Shows calls of stations recently connected to the mailbox.
JN   Shows calls of stations currently connected.

K #   Kills message number #. The message number can also be a series of numbers, (eg 112 115 117) and/or a range of numbers (eg 112 TO 115). A series can contain a range (eg 112 115 TO 117).
KM   Kills all messages addressed to you that you have read.
KT #   Kills an NTS message.

Generally lists messages in reverse order, newest to oldest. 'Private' messages not to or from you will not be listed.

L   Lists all new messages since your previous 'L' command.
LM   'List Mine'. Lists all messages addressed to you or by you.
LU   'List Unread'. Lists all messages unread by you.
L #   Lists messages back to and including number #.

LL #	Lists the last # messages.
L> call	Lists all messages to this callsign.
L< call	Lists all messages from this callsign.
L@ call	Lists all messages addressed at this BBS callsign.
LD > yymmdd	List messages newer than some date.
LD < yymmdd	List messages older than some date.
LA	List bulletins (Type A).
LB	List bulletins (Type B).
LN	List all unread messages.
LT	List all NTS traffic.
L$	Lists messages with BIDS that match a pattern. For information on patterns, do H !
LS	Lists messages with subjects that match a pattern. For information on patterns, do H !
NE	Toggle your 'expert user' status.
NH xxxxx	Enter your 'Home BBS'. (Aids in routing mesages to you.)
NN xxxxx	Enter your first name into user data base.
NZ xxxxx	Enter your zip code into user data base.
NF x	Change the format of message listings. Allowed values of x are 0 and 1.
REPLY #	Replies to msg number #.
R #	Read message number #. The message number can also be a series of numbers (eg 112 115 117) and/or a range of numbers (eg 112 TO 115). A series can contain a range (eg 112 115 TO 117).
RH #	Same as 'R' but forwarding headers are shown.
RM	'Read Mine'. Read all your unread messages.
SR xxxx	Send a reply to message 'xxxx'. You could also say REPLY xxxx.
S? xxxx @ yyy	Send message type '?' to station 'xxxx', at optional BBS 'yyy'. The mailbox will prompt for title and ask you to enter text. End text entry with a <Ctrl-Z> OR /EX. '?' is the 'type' of message.

They include:

>B – Bulletins
>P – Private. Only the addressee can read or list this type
>T – NTS traffic

T	Talk to sysop.
U topic ffff.xxx	Upload the file ffff.xxx for the topic requested. Do not use a file name that already exists.
W	Gives a list of topic areas available on the mailbox.
W topic	Gives a list of the files for that topic
W topic ffff.xxx	Gives a list of files for that topic that match the given file specification.

>W will show the file size
>WD will show the file timestamp instead of the size
>WX will include both the timestamp and size

@	Enter this symbol to indicate the BBS of the addressee, for proper forwarding of the message to its destination. The message, no matter to whom addressed, will be forwarded to the '@ BBS' location.

The G4YFB BBS mailbox software

The user commands

A*	Abort the current command being executed.
B	Bye (log off politely!)
D	Download a file from the Files section.
DU	Display the details of a user (at discretion of sysop).
H	Get the general Help File.
I	Show information about the mailbox.
J	Lists all stations connected to the mailbox.

K	Kill a message (variants at sysop's discretion).
KM	Kill Mine (kills messages to user after reading).
T	Talk to sysop (command at sysop's discretion).
U {path} [filename.ext]	Upload a file into specified section. The {path} must exist. The end of the file should be either: /ex, ^Z, ^K or /KI on a new line. The ^K or /KI aborts an upload and is not saved.
V	Version number of the BBS software.
W {directory}	What files are available in the specified directory. Also shown are any associated sub-directories.
X	Toggles the users eXpert status. Experts get a shorter prompt and no sign-off message.
?	Help. Note: help is available in the sub-directory called HELP. A user may access specified help by typing **?x** where 'x' is the first letter of the command which requires clarification.

The LIST commands

L	Lists messages down to the last one listed.
L n	List messages down to message number n.
L n m	List messages from number n to number m (can be in any order).
LA call	List messages to and from callsign (subject to sysop setup).
LL	List messages in blocks of 10 (no user prompt).
LL n	List n number of messages in blocks of 10.
LM	List all messages to user.
LN call	Lists new messages.
N	Allows new user to enter his/her name (User N or N name; <CR> leaves name unchanged).

L! TEXT STRING or text string

This command lists all those messages which contain the text string. Use lower case, ie `l! dx` will find text containing both 'dx' and 'DX'.

OS   Run specified DOS program as a separate task (sysop's DesqView installation and discretion).

R n m   Read specified message number/s (sysop's discretion).

RM   Read message/s addressed to user.

The SEND commands
S {type} {callsign} {@ BBS} {($)}

This is the general format to send a message. The {callsign} is a mandatory entry. The {$} signifies the message is a bulletin. The message should end with either: ^Z, /EX, ^K or /KI on a new line. Note: a message may be aborted by entering either ^K or /KI on a new line. The message will then be deleted.

The G1NNA mailbox software

This is listed exactly as taken from the H file. It also illustrated to this author that DXing into a mailbox is not only generally pointless but very time-consuming as well. It took three attempts, and finally over an hour, to get it early one busy evening from two nodes:

?   Type `?[LETTER]` for help with a single command. For example, type `?U` for help with uploading. Be sure to hit <Enter> or <Return> after each command.

A   Command is used to abort text download and msg list. This may not abort the list at once as there is always some text still in the buffer. Only enter 'A' once.

B   Command logs you off the BBS and disconnects your call.

D   Command lets you download files from the BBS text file section. Syntax: D [filename] <line#> <line#> You *must* use PC/MS-DOS filenames. You may specify line numbers to look at only a portion of the file, if in a DIR use `D DIRname\Filename.EXT`
Type `D USER.DOC` to download an entire file.
Type `D USER.DOC 10` to list the first 10 lines.

110   Packet Radio Primer

	Type `D USER.DOC 10 20` to list lines 10 through lines 20. Note: you can't 'download' mail. Use the 'R' command to read messages.
F	Command searches the BBS database of callsigns in order to find which BBS a given user normally uses. This information is not guaranteed to be totally accurate, since it relies on the home BBSs given when a message is sent via the NTS system.
F G9XXX	To find which BBS G9XXX normally uses.
H	Messages Commands: (K)ill (L)ist (R)ead (S)end (SR)Send Reply
	File Commands: (D)ownload (U)pload (W)hat
	Yfile Commands: (YD)ownload (YU)pload (YW)hat
	Misc Commands (H)elp (F)ind Call (V)erbose (B)ye (I)nfo (JK)Calls Connected (?)Cmd For Help, eg `?S` (T)alk to sysop
I	Command gives you information about this BBS.
JK	Command gives you a list of calls connected to the BBS.
K	Command [Kill] lets you kill only messages sent *by you* or sent *to you*. Type `K[space][message#]` to kill a specific message.
KM	Kill all messages sent to you that you have read [Kill Mine]. This will *not* kill any new mail to you that you have not read.
L	Command lists messages in the mailbox.
L	List new msgs since you last used the L command.
LM	List all msgs to you [List Mine].
LN	List unread msgs to you [List New].
LF 10	List the first 10 msgs [List First].
LL 10	List the last 10 msgs [List Last].
L< [callsign]	List msgs from [callsign].
L> [callsign]	List msgs to [callsign].
L@ [callsign]	List msgs to BBS [callsign].
L [msg#]	List only msgs *above* a given number.
L [msg#] [msg#]	List a group of messages.
L! [Word]	Searches message SUBJECT: field for a given word and returns a list of all messages containing the word, ie `L! Swap` will return

Packet mailbox commands

	a list of all messages with the word 'Swap' in the subject field. Note: check the number of *active* msgs *before* using the 'L' command.
N	N[space][YourFirstName]<Enter>' to enter your name.
R [message#]	Read a specific message. You may have up to six message numbers per line.
RM	Read all messages addressed to you [Read Mine].
RN	Read all new messages addressed to you [Read New]. To read msgs #313 and #325 type `R 313 325`. Don't forget <SPACE>.
R! [Word]	Read all messages with the [Word] in the SUBJECT: ie `R! Wanted` will allow you to read all messages with the word 'Wanted' in the subject. Use `L!` first to make sure what you are reading as there could be hundreds of messages with the word in the subject. Use with care. Note: you *read* messages sent with the 'R' commands. To *download* files you use the 'D' command.
R< [callsign]	Read msgs from [callsign].
R> [callsign]	Read msgs to [callsign].
R@ [callsign]	Read msgs to BBS[callsign].
S	Command lets you send messages (not 'files').
SP [callsign]	Send a personal message.
SP [callsign] @ [bbs]	
	Send a message to a station at another BBS by automatic mail forwarding.
SB ALL	Send a general bulletin addressed to all. Note: to upload files use the 'U' command.
SR	To send reply to a message, eg `SR 1234` will automatically a message back to the sender of No 1234.
T	Type `T<enter>` to talk to the sysop. If sysop is available to chat, you'll get a response within 30 seconds. Type anything to cancel.
U	Command lets you upload a file to the PBBS file section. Type `U filename.extension<Enter>` to upload a file. For example, type `U BORK.DOC<Enter>` to upload the file 'BORK.DOC'.

V	Shows the number of active messages and the next message number.
V <enter>	To see this BBS's software version.
V [message#]	To read a specific message. You may have up to six message numbers per line.
VM	Read all messages addressed to you [Read Mine].
VN	Read all new messages addressed to you [Read New]. To read msgs #313 and #325 type `V 313 325`.
W	Command lists the files available on disk. Depending on system capacity, files may be divided into sub-directories by class or category.
W<enter>	To list the main file directory [What files].
WN<enter>	To list new files in the main directory [What New]. 'New' means 'new since the last time you logged'. There may be sub-directories, which will appear as 'MAPS <DIR>' when you do the directory listing. To list the files in these directories, type the sub-directory name after the 'W'. For example: `W DOCS<Enter>` to list the DOCS directory or `W ICOM<Enter>` to list the ICOM directory.
Y	YAPP binary file transfer.
YW	List contents of the YAPP download directory. Works the same as W.
YN	List new entries in the YAPP download directory. Works the same as WN.
YU	YAPP upload. For example: `YU UTIL.EXE` to upload UTIL.EXE using YAPP binary transfer.
YD	YAPP download. For example: `YD FRED.COM` to download FRED.COM using YAPP binary transfer. Please note: files uploaded with YAPP go into a separate directory that cannot be accessed by users. Please let the sysop know what you have uploaded and what it does.
X	Command changes your status between 'normal' and 'eXpert' user.

The FBB BBS software

A recent addition to the ranks of BBS software is that written by F6FBB and his associates. The usage of this software is very close to the one made by WA7MBL, whose commands have been kept. It has also a set of specific and original supplementary commands.

It is a BBS bearing a close resemblance to the well-known WA7MBL or W0RLI.

Packet mailbox commands 113

It has also server functions (calculation of satellite orbits, callbook, and operator customisable chapters and gateway to an other channel, to name but a few). Note that not all of these features may be automatically implemented on your own local box. For example, the 'Gateway' is disabled on GB7BBS as this can be done by alternative methods. Those who seek information about the satellites, especially those in orbit which handle packet traffic, will see all the information they need at the press of a key. There are several features that make it resemble the sort of mailbox more often seen at the end of a 'phone line.

The forwarding is optimised (using compression techniques) between BBSs of the same type and is more efficient on a VHF network. This kind of forwarding can be disabled by configuration. The forwarding works concurrently on all channels and ports either in input or in output.

Binary transfer (of user files) is supported with the usage of the YAPP protocol of WA7MBL. An extension to this protocol has been made with the automatic restart of the protocol, should a stop have occurred or a disconnection taken place during the transfer. This extension to the protocol works with the TPK program written by FC1EBN (and should be available from your local sysop).

The menu structure is huge, as one might expect from such a mammoth program. Accordingly, only the top layer of the part seen for the first time by the user is included since the software may be configured to suit local requirements. Many of the commands are the first line in a very deep list.

A – abort	Aborts BBS output.
B – bye	Log off the BBS.
C – conference	Access to the conference.
D – DOS	Access to BBSDOS.
F – FBB	Access to server mode.
G – gateway	Access to other frequency via the 'gateway'.
H – help	Help.
I – info	Gives you information about the system.
J – jheard	Lists the last few connected stations.
K – kill	Kill messages.
L – list	List messages.
N – name	Changes your name.
O – option	Selects user options.
R – read	Read messages.
S – send	Send messages.
T – talk	Talk to sysop.
V – verbose	Verbose read of messages (including headers).
X – expert	Toggle between Normal and Expert.
Y – yapp	Yapp binary file transfer protocol.

As the configuration of the FBB software is very flexible, it is possible that the appearance may not be exactly like the above, nor is it always necessary that all the facilities are implemented, since it is often up to the local sysop and his users to specify what they want. This makes it difficult to document, but there are four user guides which may be downloaded and read at leisure.

One interesting feature is that it will run (if suitably configured) in 11 languages and may automatically respond in the language assigned to the connected callsign, so don't be too surprised if you see German or Spanish called as the result of a visitor logging in!

Appendix 3

A short guide to communication cables

There is often much confusion in the amateur (and the wider) world about serial communications. RS232 is not the easiest thing to find out about. In fact, since the latest revision, it isn't even called RS-232-C any more. RS-232-C was a standard recommended by the Electronic Industries Association in 1969, since which date the development of communications has leapt forward. The latest version at the time of writing is now called EIA-232-D (published in 1987), which makes it more compatible with the recommendations of the CCITT specifications V24 and V28, as well as ISO IS2110.

We are still dealing here with the method of connecting a data terminal equipment (DTE) (a terminal or computer behaving like one), which is fitted with a plug, and a data communication equipment (DCE) (like a TNC or a modem), which has a socket. As far as can be found, Table 7 is the details of all the connections specified.

To look at Table 7, you would think that someone could have put a few more pins into the plug and done the lot, as opposed to the variations permitted for pins 21 and 23. Fortunately, these are not often seen in amateur equipment, so we can happily ignore them. Some computers have a 9-pin DB type connector, but the signals we need are all there.

As far as we are concerned, the general steps in communications are as follows (remember that the DTE is the Terminal or Computer, and the DCE is the TNC):

DSR (DCE Ready)	DCE tells DTE it is prepared (set up and on).
DTR (DTE Ready)	DTE tells DCE it is prepared.
RSLD (DCD)	When my TNC is connected to another station this line goes true.

Once connected and set up these lines often stay that way. But to continue:

RTS	DTE tells DCE it has data to send.
CTS	DCE tells DTE: OK, get on with it.
Transmit data	The data which the DCE will send into the medium (phone line or radio).
Receive data	Demodulated data from the DCE.

Table 7. RS232 cable connections

Pin	EIA	CCITT	Name	Also known as
1	–	–	Shield	Frame Earth
2	BA	103	Transmitted Data	TXD
3	BB	104	Received Data	RXD
4	CA	105	Req to Send	RTS
5	CB	106	Clear to Send	CTS
6	CC	107	Data Set Ready	DCE Ready [DSR]
7	AB	102	Signal Earth	Signal Common
8	CF	109	Rec Sig Line Det	Carrier Det [DCD]
9			Reserved for Test +ve	Test Volts
10			Reserved for Test -ve	Test Volts
11			Unassigned	Select TX Freq.
12	SCF	122	2nd RX Line Sig Det	Back chan RLSD
13	SCB	121	2nd CTS Back chan	CTS
14	SBA	118	2nd TXD Back chan	TXD2
15	DB	114	TX Timing (from DCE)	TX Clock
16	SBB	119	2nd RXD Back Chan	RXD2
17	DD	115	RX Timing	RX Clock
18	LL	141	Local Loopback	LL [not on 232-C]
19	SCA	120	2nd RTS Back chan	RTS
20	CD	108.2	Data Term Rdy	DTE Ready [DTR]
21	CG	110	Sig Qual. Det.	SQD
also:	RL	140	Remote Loopback	SQD or RL
22	CE	125	Ring Det Ring Ind	RI
23	CH/CI	111/112	Data Signal rate	Det DRS
24	DA	113	TX Timing (from DTE)	Ext TX Clock
25	TM	142	Test Mode	Test Ind

Connecting things up

Depending upon the particular TNC you choose, you may require to make up a variety of connecting leads (this actually goes for connecting a wider variety of

Table 8. Connections to a PC serial port

9-pin	25-pin	Mnemonic	Name
shell	1	FG	Frame Ground
3	2	TXD	Transmitted Data
2	3	RXD	Received Data
7	4	RTS	Request To Send
8	5	CTS	Clear To Send
6	6	DSR	Data Set Ready
5	7	SG	Signal Ground
1	8	DCD	Data Carrier Detect
4	20	DTR	Data Terminal Ready
9	22	RI	Ring Indicator

Note: the above names are those given in most books on RS232, rather than those for V24 or CCITT specs which may have other nomenclatures. See Table 7 for other names.

A short guide to communication cables

Table 9. TNC 200/220 signal connections

Pin	Mnemonic	Name	Direction
1	FG	Frame Ground	
2	TXD	Transmit Data	I/P to TNC
3	RXD	Receive Data	O/P from TNC
5	CTS	Clear To Send	O/P from TNC
6	DSR	Data Set Ready	O/P from TNC
7	SG	Signal Earth	
8	DCD	Data Carrier Detect	O/P from TNC
20	DTR	Data Terminal Ready	I/P to TNC
23	RI	Ring Indicator	not often used

DCEs to your computer). In practice, different software may require different methods of connection. The following 'null modem' connections are included for completeness, as you may find them useful. Our thanks to INMAC (UK) Ltd, for their permission to quote from their catalogue.

A PC or PC/AT type computer will require connections to the pins on its serial port given in Table 8.

The TNC 200/220 has the signals given in Table 9 on the pins of the 25-way D-type connector.

Appendix 4

Some PC software

YAPP, v2.0

YAPP (Yet Another Packet Program) was written by Jeff Jacobsen, WA7MBL, who wrote the Bulletin Board System, and it is designed for use on the IBM or compatible computers PC, XT and AT. In order to transfer and use binary files, it requires hardware handshaking, ie eight of the pins on an IBM-type nine-pin D-type plug are used; (on the 25-pin connection, use pins 1-8 and pin 20). There are in addition some settings to be made on the software side of the commands set to the TNC by the YAPP.CNF configuration file:

```
START 0
STOP 0
XON 0
XOFF 0
XFLOW OFF
```

Other software may require these set to their respective default settings. Where split-screen mode is used, it may be beneficial to set the following in addition to those above:

```
FLOW OFF
ECHO OFF
```

The configuration file may be amended or written using any text processor program (like WordStar in NON-DOC mode). It is, however, important to read the YAPP.DOC file thoroughly, as a lot of confusion can easily arise in getting the settings incorrect. An example of the .CNF (configuration) file used at G8UYZ might be instructive, as it shows the whole of the configuration:

```
1       {Comms port}
4800    {Baud rate}
2       {TNC type}
NO      {split screen
```

```
YES    {allow bells (^G)}
6      {status line attribute}
2      {standard attribute - keyboard input}
14     {received character attribute}
7      {help screen attribute}
4      {help screen border}
15     {transfer status attribute}
4      {transfer status border}
15     {set default window attribute}
4      {set default window border}
15     {directory window attribute}
4      {directory window border}
15     {misc window attribute}
4      {misc window border}
```
awlen 8
parity 0
restart
ax25 on
beacon every 60
btext Wolverhampton BBS is off-line
cbell on
cmsgd off
conok on
constamp off
dayusa off
delete off
dwait 4
digi on
flow on
headerln on
hid on
mycall g1dil
mrpt on
mcon off
mstamp off
maxframe 4
screenl 0
start 0
stop 0
streamcall on
txdelay 20
txflow off
xflow off
xoff 0
xon 0
unproto beacon via AP
users 1
monitor on

```
    mheard
    *** EOF  {end of commands sent to TNC on program start}
    mon off
    conok off
    *** EOF  {end of commands sent to TNC on program end}
```

It is possible to set up the colours used on screen. The following is a copy of a helpful file (COLOR.DOC), that came with my copy of YAPP:

```
For those of you who were trying to figure out what colour 111 is or how to
change the Help background to something else, here is a chart to guide you
through choosing your colours for YAPP. I hope it helps you out.

73, Mark N2MH

Monochrome Values:

  0  -  Nondisplay
  1  -  Underline
  7  -  Normal video
  9  -  High intensity, underline
 15  -  High intensity
112  -  Reverse video

Other values will produce one of the above attributes. For blinking
characters, add 128 to the values given for monochrome or for colour.

The colour chart is shown below
Background   |       |      |       |      |     |        |       |       |
Colours  =>  | Black | Blue | Green | Cyan | Red |Magenta| Brown | White |
-------+---+---+---+---+---+---+---+---|
Foreground   |       |      |       |      |     |        |       |       |
Colours      |       |      |       |      |     |        |       |       |
   |         |       |      |       |      |     |        |       |       |
   |         |       |      |       |      |     |        |       |       |
             |       |      |       |      |     |        |       |       |
-------+---+---+---+---+---+---+---+---|
Black        |   0   |  16  |  32   |  48  |  64 |   80   |  96   |  112  |
Blue         |   1   |  17  |  33   |  49  |  65 |   81   |  97   |  113  |
Green        |   2   |  18  |  34   |  50  |  66 |   82   |  98   |  114  |
Cyan         |   3   |  19  |  35   |  51  |  67 |   83   |  99   |  115  |
-------+---+---+---+---+---+---+---+---|
Red          |   4   |  20  |  36   |  52  |  68 |   84   | 100   |  116  |
Magenta      |   5   |  21  |  37   |  53  |  69 |   85   | 101   |  117  |
Brown        |   6   |  22  |  38   |  54  |  70 |   86   | 102   |  118  |
White        |   7   |  23  |  39   |  55  |  71 |   87   | 103   |  119  |
-------+---+---+---+---+---+---+---+---|
```

```
Gray            |  8 | 24 | 40 | 56 | 72 | 88 | 104 | 120 |
Light Blue      |  9 | 25 | 41 | 57 | 73 | 89 | 105 | 121 |
Light Green     | 10 | 26 | 42 | 58 | 74 | 90 | 106 | 122 |
Light Cyan      | 11 | 27 | 43 | 59 | 75 | 91 | 107 | 123 |
----+----+----+----+----+----+----|
Light Red       | 12 | 28 | 44 | 60 | 76 | 92 | 108 | 124 |
Light Magenta   | 13 | 29 | 45 | 61 | 77 | 93 | 109 | 125 |
Yellow          | 14 | 30 | 46 | 62 | 78 | 94 | 110 | 126 |
Hi. Int. White  | 15 | 31 | 47 | 63 | 79 | 95 | 111 | 127 |
```

ProComm, v2.41

Specifically for the PC, this is a shareware piece of software (from DataStorm Technologies) which is easy to use, comprehensive, and very flexible when it comes to setting up the parameters controlling how it will behave. A more recently seen version is ProComm Plus, which does the same thing but offers a few other facilities. Everything is set up from within the program using a very good menu system and, once configured, all parameters may be saved on the disk so that future use will not require any changes.

It supports many varieties of file transfer modes, will emulate many popular terminals, and has a 'chat mode' (split screen), as well as some limited host mailbox facilities. It has been well proven in many telecommunication situations and will often work when others stubbornly refuse. For packet, the one thing missing is support of the YAPP protocol for file transfer. Later versions are available as shareware (try before you buy), and any users are advised to register their copy, as the manual is quite large.

ProComm will allow you to do many operations at the touch of a single key (macros or scripts), and this is found by many to be very handy. Below is the SETTINGS.CMD file used at *chez* G8UYZ. Note that comments are preceded by a semicolon (;). It's actually rather like REM in BASIC.

Thanks are due to G1NQW who put G8UYZ in better command of the TNC.

Comments have been added for some clarity. The RESTART at the end is important, particularly if you have changed the bit rate and parity. If you do this sort of thing regularly, put that section at the end of the file, or else some commands may be ignored (the TNC won't understand them properly). The exclamation mark (!) at the end of each command in the quotes is ProComm-speak for a carriage return.

```
; TNC Set -up script file:    settings.cmd
; to summon and impliment default values for my TNC200, v1.1.6B2
; G8UYZ, 1991.
;
transmit "LFI ON !"       ; Ignore Line Feeds for this file
transmit "XM ON!"         ; Transmit permission granted -
;                           very useful with youngsters and relatives
```

```
;
; Communication Commands:-
transmit "AW  8!"        ; 8 bits
transmit "PAR 3!"        ; parity
transmit "AU  OFF!       ; AUto line feeds
transmit "8b OFF !"      ; no 8th bit strip
transmit "F ON!"         ; Flow , typing over-rides input frame
transmit "BR OFF!"       ; Break character ignored
transmit "LC ON!"        ; Case conversion
transmit "NU OFF!"       ; No nulls <CR>
transmit "NUL OFF!"      ; No nulls <LF>
transmit "NULL 0!"       ; No nulls <both>
transmit "TRF ON!"       ; TRFlow: software handshake
transmit "TXF ON!"       ; TXFlow:   -"-
transmit "XFLOW OFF!"    ; Hardware Flow control
;
; Character Commands:-
transmit "BK ON!"        ; Backspace on Delete,
transmit "DEL OFF!"      ; not the delete key
transmit "COM $03!"      ; Command character
transmit "LCS OFF!"      ; LowerCase Streamswitch
transmit "STA $11!"      ; Start
transmit "STO $13!"      ; Stop
transmit "STREAMC ON!"   ; Callsign after streamswitch
transmit "STREAMD OFF!"  ; Double stream sign
transmit "STR $7C!"      ; Stream Switch character
transmit "XOFF $13!"     ; ^S (FLow Control)
transmit "XON $11!"      ; ^Q ( -""- )
;
; PMS Commands:-
transmit "3rd OFF !"     ; No third Party
transmit "KIS OFF!"      ; No KISS
transmit "PMS ON!"       ; Answering machine on
transmit "HOM GB7BBS-15" ; Home BBS for messages
transmit "KILONFWD ON!"  ; Wipe mesage on Transmit from PMS
transmit "LOG OFF!"      ; Supress PMS LogOn message, use SText
transmit "MSGH ON!"      ; Forwarded messages have headers
transmit "MYA G8UYZ-3!"  ; My ALias
transmit "MY G8UYZ!"     ; ME
transmit "MYP G8UYZ-2!"  ; PMS Callsign
;
; On screen display:-
transmit "AD  ON !"      ; Monitored Frames with Address
transmit "AM  ON!"       ; Alphabetic Month display
transmit "Echo OFF!"     ; No double characters
transmit "DAYU OFF!"     ; European format date
transmit "S 80!"         ; 80 chars on 1 line
;
```

```
; Link Commands:-
transmit "CONM C!"       ; Converse mode on connection
transmit "CONOK ON!"     ; Allow Conections
transmit "CONP OFF!"     ; Temporary connection
transmit "DIG OFF!"      ; No digi-peating
transmit "FU OFF!"       ; Half Duplex (FUll)
transmit "LF OFF!"       ; Line Feeds
transmit "MAX 2!"        ; 2 Frames outstanding
transmit "NE ON!"        ; Mode change automatic
transmit "NO OFF!"       ; No mode change
transmit "US 1!"         ; One user
;
; Monitor Commands:-
transmit "CB ON!"        ; ^G Bell enabled
transmit "CONS OFF!"     ; Don't time-stamp the packets
transmit "CP OFF!"       ; No timed despatch (CONV mode)
transmit "CR ON!"        ; Append <CR> to each packet
transmit "CRA ON!"       ; Append <CR> after monitored frame
transmit "HE ON!"        ; Header on separate line
transmit "HI OFF!"       ; Disable ID packet
transmit "MA ON!"        ; Monitor All frames
transmit "MCOM OFF!"     ; Data Frames only
transmit "MCON OFF!"     ; No other frames when connected
transmit "Mon ON!"       ; Monitor
transmit "MR ON!"        ; Display digi path with frames
transmit "PI ON!"        ; Pid check: nodal garbage filter
;
; Timing Commands :-
transmit "AXD 2!"        ; AXDelay 200 mSec.
transmit "DW 8!"         ; Extra delay for digi's
transmit "FR 4!"         ; Frame Ack wait time
transmit "CH 30!"        ; Connection Timer
transmit "CM 1!"         ; Timeout value
transmit "P 128!"        ; Packet Length
transmit "PE 127!"       ; Persist, retry probability
transmit "PP OFF!"       ; Persistance or Frack timing
transmit "RES 1!"        ; Ack timing
transmit "RE 10!"        ; Retries
transmit "SE $0D!"       ; SENDPAC character ( = <Enter> key)
transmit "SL 1!"         ; Slot time ( for Persist timing)
transmit "TRI 0!"        ; Retry counter
transmit "TX 30!"        ; TX Delay 300 mSec
;
; Misc Commands
transmit "RED $12!"      ; Redisplay line character
transmit "RX OFF!"       ; Blocks for Computer processing
transmit "TRAC OFF!"     ; Trace mode
;
```

```
; Station IDent commands:-
transmit "CWid A 18!"     ; CW ident every 15 min.
transmit "CWL 6!"         ; @ 20wpm (max speed allowed)
transmit "CWIDT G8UYZ!"   ; & what is sent
transmit "CMS ON!"        ; Connect message enabled
transmit "CMSGD OFF!"     ; Don't force disconnection
transmit "U CQ!"          ; Unproto CQ
transmit "B A 90!"        ; Beacon after activity
transmit "CT  Please leave a message on G8UYZ-2. Thank you: 73,
Dave.>!"
transmit "STE  Welcome to the G8UYZ-2  Answering machine.!"
transmit "BText G8UYZ @ Penkridge, IO82WR:  WAB SP91,
Staffordshire.!"
transmit "LFI OFF!"       ; Back to square 1
transmit "restart!"       ; Reset to beginning and ready
exit
```

The `exit` at the end transfers control back to the TNC after the commands are entered. Note that the file above uses Software FLow control (G8UYZ was suffering problems at the time it was written). There may seem to be a lot of commands, but the list is not actually complete. For your own use, you need only use those entries which are different from your TNC's default settings.

Please remember that RESET and RESTART are not the same thing. RESET will change all your settings back to factory default (ie MYcall = NOCALL), whereas RESTART will re-boot the machine as if you'd just turned it on.

The settings on the Timer section are for the FT221RD at station G8UYZ. You will obviously have to change these and any others to your own requirements in your own station.

To use the file, press Alt-<F5> and select which of the scripts you want to use with the arrow keys, and press <Enter>. If you want to do this sort of thing with ProComm Plus, change the file name to SETTINGS.ASP.

One thing to remember – if it works, leave it alone!

It is important to note that later versions of ProComm, such as ProComm Plus, have similar facilities, but are used in a different way. Please see the appropriate manual for further information. If you use it you should register your use with Datastorm or their agents.

A brief look at Paket, v4.0

Paket4 is a recently-discovered program for the PC which has a lot of useful features for the packet enthusiast. Those who have tried to send binary files from one station to another may well find that it is even better. However, it is not the purpose of this piece to advertise or reccomend, so we will just suggest a couple of useful starting points for its use.

It is important that Paket4 sees a clean hardware handshake. Station G8UYZ suffered a long time trying to get it right on the old TNC-200 but after many attempts, files are now being sent at a reasonable average speed of 90 bytes/sec, compared with the maximum theoretical rate of 120 bytes/sec. If Paket4 does not see a clean connection, it resets the mode and tries to use software handshaking (XON/XOFF). This can slow things down a good deal. Software handshaking has worked successfully – but do check that it is enabled at both TNC and within the Paket4 configuration file.

The average PC does not have RTS on the serial port. It follows, then, that careful attention is paid to the connections, as CTS will be doing all the work, whilst DSR and DTR are firmly wired and working. G8UYZ has all available wires connected, and lets the PC and TNC do what they will.

The asynch parameters at G8UYZ are currently set thus:

```
8bitconv   ON
AUtolf     OFF
AWlen      8
BBSmsgs    OFF
BReak      ON
Echo       OFF
EScape     OFF
Flow       ON
LCok       ON
NUcr       OFF
NULf       OFF
NULLs      0
PARity     0
RXblock    OFF
Screenln   80
TRFlow     OFF
TXFlow     OFF
Xflow      OFF
```

The settings of the timers can be a matter for much debate. FRACK, MAXFRAME and PACLEN all have a significant bearing on the effective transfer rate. Attempts between G8UYZ, G8ZWU and G0NEN seem to indicate that FRack = 5, MAXframe = 6 and PAClen = 228 works quite well, although experiments continue.

```
cmd:disp t
AXDelay    1
AXHang     0
CALSet     0
CHeck      30
CLKadj     0
CMdtime    1
CPactime   OFF
DWait      12
```

```
FRack      5
PACTime    AFTER 10
PErsist    127
PPersist   OFF
RESptime   0
SLottime   1
TXdelay    40
```

The TX and AX delays are included for completion, but the old radio to which the TNC is connected (an FT780), requires a bit of time to lock up properly in transmit mode, so they are a bit longer than the defaults.

The Paket4 manual is a little difficult to follow, but it is worth the effort.

The default setting of PACLEN for YAPP parameter in Paket4 is 0, or 256 bytes. This causes problems with PacComm 1.1.6C firmware, and should be changed: 255 works, but 128 or so is better on a busy channel. We don't know why the program was distributed like this. It affects a large proportion of potential users but is only mentioned in the small print deep in the bowels of the manual – just begging for adverse reviews!

In using the REMOTE mode, it is important to type the filename as well as the command, ie YU filename, rather than wait for it to prompt you to give it later. If you have to use the DOS-gateway (F9), do not be alarmed at the curious prompt. It works anyway, although you may have to be careful about typing in a new path. When leaving the REMOTE mode, we have found it necessary to send a ^C to force the return of the command prompt. Don't forget that Paket4 sends a ^Z at the end of an ASCII file, so you'll not need to do it manually.

There will be more of these discoveries as knowledge grows; but keep trying – the program *is* worth the effort.

As Murphy once put it: "If all else fails, RTFM".

Appendix 5

A short glossary

Archive
A method of storing a file or group of files in such a manner as to take less space on the disk. Reductions in space of between 10 and 90% are possible. It is used for files which are not in use, but are themselves useful to someone. The compression utility should be accompanied by the corresponding expansion utility. It is also possible to make a file expand automatically. Examples are ARC, by SEA, and PK (PKARC, PKXARC PKZIP and PKUNZIP (which are later versions of PKARC) by PKWare which are frequently seen on mailboxes as well as in the catalogues of shareware dealers). Some code the compressed files with a Q in the centre of the extension (.DOC = .DQC).

A group of files may be stored as a collective .ARC or .ZIP file. An individually 'ARCed' file may sometimes be seen as having an unexpected middle letter of the extension, ie .DQC in place of .DOC, or .TQT instead of .TXT. Such files are also seen in custom installation programs of commercial software.

Binary file
A file containing more than ASCII text which includes control codes, such as 'actual programs' which are in a form which may be executed or run.

Baud rate
A baud rate of 1 is one signal element per second. A baud rate of 1200 indicates that transmission speed is about 120 letters or numerals per second. It is named after the French telegraph engineer J M E Baudot. Some of the later speeds are measured in 'bits per second'.

Compression
See *Archive*.

Digipeater
A station capable of repeating the AX25 frames. Usually operated 'transparent' to the operator of the digi station. See also *Nodes*.

Hardware
The integrated circuits and their connections, together with any other items (power supplies, disk drives, etc) which make up the device or unit.

Handshaking
The process whereby the TNC and its computer or terminal exchange acknowledgment of each others' existence and preparedness to send and receive traffic. This may take the form of control codes sent with the data (software handshaking) or on lines separate from those carrying the data (hardware handshaking). Hardware handshaking is used by some software in order that binary files (ie executable programs etc) are not corrupted by control codes which may mean something to the program being transferred.

Software handshaking is used for most applications like the transfer of text files and normal messages (which are ASCII files). You will probably use the commands XON and XOFF in cases where software handshaking is used.

Modem (MODulator/DEModulator)
Creates the tones used in the transmission of data down a telephone line, or, in our case, the tones transmitted.

Node
A method of 'repeating' the packet data from one point to another in such a manner as to make it apparently faster and with less visible errors. The real difference is more fundamental; acknowledgements are sent between nodes rather than from one node to the other (as is the case with multi-hop digipeating). See Fig 10.

Software
The instructions or program routines which cause a desired action in the hardware. It is usually found as either in ROM or loaded from some exterior source (eg disks or ports) and put into RAM, from where it is actioned.

Wildcard
On many computers it is possible to put in a wildcard (which often takes the form of a '*' or sometimes one or more '?' characters) when requesting files of a particular type. Thus, to list all the README files for a subject, when you don't know if they are .TXT or .DOC files you could type

```
LIST README.*
```

and all the files which have README.something will be listed. The '?' may sometimes replace the '*' when used. For example, some files are archived (a storage method which save much disk space), so you could type

```
LIST README.D?C
```

which will list all the .DOC as well as any .DQC (archived) files.

Appendix 6

Guidelines for the use of the packet radio network

The RSGB Data Communications Committee, in consultation with the Radiocommunications Agency, has devised the following guidelines for the use of the UK packet radio network, with which all operators are urged to comply.

These guidelines have been split into four sections in order to reflect:

A. The need for messages to be within the terms of the licence conditions and the implications if they are not.
B. Messages which could result in legal action being taken by other amateurs or outside bodies.
C. Actions to be taken when amateurs identify cases of abuse.
D. Other appropriate items.

Section A

1. All messages should reflect the purposes of the amateur licence, in particular 'self-training in the use of communications by wireless telegraphy'.
2. Any messages which clearly infringe licence conditions could result in prosecution, or revocation, or variation of a licence. The Secretary of State has the power to vary or revoke licences if an amateur's actions call into question whether he is a fit and proper person to hold an amateur licence. An example of this could be unreasonable behaviour by using the packet network to carry on a dispute or to deliberately antagonise other amateurs.
3. The Radiocommunications Agency has advised that the Amateur Radio Licence prohibits any form of advertising, whether money is involved or not.
4. Messages broadcast to 'ALL' stations are considered acceptable but should only be used when of real value in order to avoid overloading the network.
5. Do not send anything which could be interpreted as being for the purpose of business or propaganda. This includes messages of, or on behalf of, any social, political, religious or commercial organisation. However, our licence specifically allows news of activities of non-profit-making organisations formed for the purposes of amateur radio.

Section B

1. Do not send any message which is libellous, defamatory, racist or abusive.
2. Do not infringe any copyright or contravene the Data Protection Act.
3. Do not publish any information which infringes personal or corporate privacy, eg ex-directory telephone numbers or addresses withheld from the *Call Book*.

Section C

1. Any cases of abuse noted should be referred in the first instance to AROS* which will take the appropriate action.
2. It is worth noting that any transmissions which are considered grossly offensive, indecent, obscene or menacing should be dealt with by the police. This action should also be coordinated by AROS* initially.
3. Mailbox sysops have been reminded by the Radiocommunications Agency that they have an obligation to review messages daily and that they should not hesitate to delete those that they deem unacceptable. It is worth remembering that their licence is also at risk as well as your own.

Section D

1. Do not send 'open letters' to individuals.
2. Do not write in the heat of the moment. Wordprocess your bulletin first, then re-read it. You may feel differently after a few minutes.
3. Obey the Golden Rule – if you would not say it on voice do not send it on packet.

*The RSGB's Amateur Radio Observation Service (see the RSGB *Amateur Radio Call Book* or *Radio Communication* December 1988).

Index

A

ALOHANET, 3
Amateur Radio Observation Service, 132
Archive, 129
ASCII code, 10
AX25, 3

B

Baud rates, 31
    definition, 129
    radio port, 15
    serial port, 20, 21
Beacon text, 39
Bell, use of, 40, 45, 64
Binary numbers, 10
Bulletin boards, 8, 51
Bulletins, 90

C

Cables, communication, 115
Callsign identification, setting up on TNC, 37, 39
Chevron symbol, use of, 44
Clusters, 74
Command mode, 41
Computers
    connections, 21, 27, 117
    principles, 8
    requirements for packet, 18
Connect text, 41
Connections
    radio port, 29
    serial port, 21, 27, 117
Connectors, 21, 27
Conversation mode, 41
CPU (central processing unit), 9

D

Data
    links, 21, 25
    principles, 10
Date, setting up on TNC, 43
DCE (data communications equipment), 25, 115
Digipeaters, 8, 46, 129
DTE (data terminal equipment), 25, 115
DX PacketClusters, 74

E

EIA-232-D, 25, 115
EMC problems, 30

F

Files
    binary, 129
    compression of, 129
    downloading, 56, 90, 91
    uploading, 95, 97

G

G-bell, use of, 40, 45, 64

H

Handshaking, 25, 130
Hardware, 130
Hexadecimal numbers, 10

I

Interference, 30

L

Line monitor, 26
Logs, 43

M

Mailboxes, 51
   AA4RE, 104
   commands, 89
   FBB, 112
   G1NNA, 109
   G4YFB, 107
   personal, 56
   W0RLI, 55, 99
   WA7MBL, 53, 89
Mark tone, 10
Modems, 11, 130

N

NET/ROM, 8, 69
Nodes, 47, 69, 130
   route table, 70, 72

O

Operating guidelines, 44, 131

P

Packets, 7
   frames, 67
   length, 42
   protocol, 67
Paket4, 125
Performance adjustments, 44
PMS (personal message system), 56, 61
   as sysop, 61
ProComm, 122

R

Radios
   connections, 29
   requirements for packet, 15, 44
RAM (random-access memory), 9
ROM (read-only memory), 9
   v1.1.6 in TNC, 60

Route qualifier, 90
Route table, node, 70, 72
RS232, 21, 25, 115

S

Serial port, 19, 21, 25
Software
   principles, 9
   programs available, 130
   requirements for packet, 20
Space tone, 10
SSID (secondary station identifier), 61, 67
Station
   callsign, setting up, 37, 39
   requirements for packet, 15

T

TAPR (Tucson Amateur Packet Radio), 3
Terminals
   emulation, 20
   requirements for packet, 16
TheNET, 8, 69
Time, setting up on TNC, 43
TNC (terminal node controller), 7, 21
   commands, 81
   connections, 25, 29
   parameters, 31, 37
   requirements for packet, 16
   ROM v1.1.6, 60
   setting up, 31, 37
   Tiny-2, 27, 29
   TNC200, 28, 117

W

Wildcard character, 130

Y

YAPP, 97, 119

Get more out of amateur radio ...
... as an RSGB member!

❏ BOOKS
The Society publishes a wide range of books on amateur radio at reasonable prices. In addition, members enjoy a discount on all titles.

❏ CONTESTS
Several contests are organised each year. It's a fun way to test your operating skills and, in the case of club contests, improve your social life!

❏ EQUIPMENT INSURANCE
Insurance for your valuable equipment has been arranged specially for members at advantageous rates.

❏ OPERATING AWARDS
A wide range of distinctive awards is available to help you set operating goals and reward your achievements.

❏ PLANNING PERMISSION
There is a special booklet and expert help available to members seeking assistance with planning matters.

❏ QSL BUREAU
This invaluable service sends and receives your QSL cards on a world-wide basis.

❏ RADIO COMMUNICATION
The leading British radio amateurs' magazine, published monthly. It's packed with technical features, regular columns (including one for datacomms) and the latest news.

Send for our Information Pack and discover how you too can get more out of amateur radio. Write to:

RADIO SOCIETY OF GREAT BRITAIN
Lambda House, Cranborne Road,
Potters Bar, Herts EN6 3JE

Some other RSGB publications...

❑ AMATEUR RADIO CALL BOOK
As well as a list of all UK and Republic of Ireland radio amateurs, this essential reference work also includes an information directory giving useful addresses, EMC advice, lists of amateur radio clubs, operating data, and much more.

❑ AMATEUR RADIO OPERATING MANUAL
Covers the essential operating techniques required for most aspects of amateur radio including DX, contests and mobile operation, and features a comprehensive set of operating aids.

❑ MICROWAVE HANDBOOK
A major new publication in three volumes. Volume 1 covers operating techniques, system analysis and propagation, antennas, transmission lines and components, semiconductors and valves. Volume 2 continues with construction techniques, common equipment, beacons and repeaters, test equipment, safety, filters and data. Volume 3 concludes with practical equipment designs for each band.

❑ RADIO AURORAS
This new book gives a technical account of the latest research into how auroras are caused, how they can be forecast, and how best to use them to work DX.

❑ SPACE RADIO HANDBOOK
Space exploration by radio is exciting and it is open to anyone! This new book shows you how it is done and the equipment you will need. It covers the whole field of space radio communication and experimentation, including auroras, meteor scatter, moonbounce, satellites, manned spacecraft and simple radio astronomy.

❑ VHF/UHF MANUAL
This standard textbook on the theory and practice of amateur radio reception and transmission between 30MHz and 24GHz includes full constructional details of many items of equipment.

RADIO SOCIETY OF GREAT BRITAIN
Lambda House, Cranborne Road,
Potters Bar, Herts EN6 3JE

PACKET RADIO PRIMER (1st edn)

We hope you found this book interesting and useful. Please let us have your comments and suggestions for the next edition so we can make it even better!

CUT ALONG DOTTED LINE

Name... Callsign........................

Address...

....

....

....

FOLD 1

AFFIX
STAMP
HERE

RSGB Book Editor
Radio Society of Great Britain
Lambda House
Cranborne Road
POTTERS BAR
Herts EN6 3JE

FOLD 2

SEAL WITH ADHESIVE TAPE HERE